The Artificial Ear

Cochlear Implants and the Culture of Deafness

STUART BLUME

RUTGERS UNIVERSITY PRESS
New Brunswick, New Jersey, and London

Library of Congress Cataloging-in-Publication Data

Blume, Stuart S., 1942–
The artificial ear : cochlear implants and the culture of deafness / Stuart Blume.
p. cm.
Includes bibliographical references and index.
ISBN 978–0–8135–4659–9 (hardcover : alk. paper) — ISBN 978–0–8135–4660–5 (pbk. : alk. paper)
1. Cochlear implants—Social aspects. 2. Cochlear implants—History. I. Title.
RF305.B58 2010
617.8'8220592—dc22 2009016194

A British Cataloging-in-Publication record for this book is available from the British Library

Visit our Web site: http://rutgerspress.rutgers.edu

Manufactured in the United States of America

For Jascha and Boaz

Contents

Preface

This book has been long in the making and has deep roots. My interest in the development and the consequences of new technologies of health care was the result of my work, in the 1970s, as secretary of a British government committee on Social Inequalities in Health. How and why innovation in health care technology seems so often to exacerbate social inequalities, to be the cause of ethical, social, and economic problems in addition to providing cures, is a question that has preoccupied me ever since. It has roots in personal experience too. In 1989, my son Jascha, born in 1987, was found to be deaf. His brother Boaz, born in 1992, is hard of hearing. What did medical technology have to offer them? I decided that I would spend some time on a study that in one way or another would help me understand. How and why this became a study of the cochlear implant is discussed in the first chapter of this book. When I started work, more than fifteen years ago, I had little sense of how long it would take or how difficult it would be.

Early in 1999, on vacation in Switzerland, I read Arthur W. Frank's wonderful book *The Wounded Storyteller.* The insights that it provided in the relations between suffering, coping, and storytelling came as a revelation. If I had not read Frank's book at that time I do not know if this one would ever have been completed. I read *The Wounded Storyteller* at a time in which I seemed no longer able to work on my study of the cochlear implant. I had begun a quite different, unrelated project, and returning to this one seemed too painful and too difficult. Slowly, and thanks to Frank's book, I started to understand why this was so, but also to understand why, despite everything, I needed to finish my own.

My difficulties had to do with the fact that the study was part of my attempt to come to terms with the place that deafness now had in my life, and that that "coming to terms" had ended. I had come to see my children no longer as "deaf" but simply as "my children." Frank's book showed me that not only was I too a "wounded storyteller" but that the storyteller has responsibilities not only to him- or herself but to others who might be helped or inspired by what he or she has to say.

In the course of so many years, I have accumulated more debts than I can acknowledge here. I would, however, like to express my particular thanks to František Bouška, David Brien, Rita Bruning, Jean Dagron, Liisa Kauppinen, Harlan Lane, Bernard Mottez, Gunilla Őhngren, Olivier Perrier, Johan Ros, Terry Shinn, Anneke Vermeulen, Elizabeth Vroom, Johan Wesemann, and Lucy Yardley for their collaboration, advice, and support. The Wellcome Trust Programme in the History of Medicine provided financial support for a part of the research, while a fellowship from the French National Research Council (CNRS) and the hospitality of the GEMAS/Maison des Sciences de l'Homme made it possible for me to spend a number of fruitful months in Paris. Isabelle Baszanger, Rayna Rapp, Ulrike Lindner, and Maria Fernanda Olarte Sierra provided invaluable comments on versions of the manuscript, in whole or in part. Sib de Boer and Annette Portegies encouraged me to complete and publish a Dutch account of my experiences, while my editor at Rutgers, Doreen Valentine, has been a wonderful source of encouragement and advice. To all of them, and to Anja with whom I have shared so much of the experience that helped shape this book, I am deeply grateful.

The Artificial Ear

Chapter 1

The Promise of New Medical Technology

Impaired hearing is widespread. An estimated 22 million Americans suffer from it, of whom more than 10 million have difficulty following normal conversational speech. Although hearing loss is far more common in older people, it also affects close to half a million children in the United States. The vast majority of these children, even those born totally deaf, have parents who can hear. Confronted with the diagnosis that their child is deaf, parents are shocked and bewildered. Only gradually, as shock and (not uncommonly) denial that anything is wrong have been overcome, can the complex business of coping with the implications of having a deaf child in the family begin.[1]

In early 1989, concerned that there might be something wrong with our young son Jascha's hearing, my wife Anja and I consulted a specialist in ear diseases. Finding nothing organically wrong with Jascha's ears, he suggested we visit an institute for pediatric audiology near where we live, in Amsterdam, the Netherlands. The Dutch Foundation for the Deaf and Hard of Hearing Child, or NSDSK, would be able to provide us with some answers.

The audiologist asked Anja to sit in a chair at the center of the room, surrounded by screens, monitors, and toys. Jascha had to be kept calm and alert, and sat on her lap. I was told to sit in the corner. I could watch. This I did with total absorption as the audiologist and his assistant carried out the test. The audiologist sat at the console, gradually filling in values on the computer. As Jascha responded, or failed to respond, to sounds of different frequency and loudness, an entry was made. Slowly a picture of Jascha's hearing was built up on that computer. Neither of us was prepared for what the audiologist had to say, his assessment complete: "I'm afraid I have bad news for you. Your son is profoundly deaf. You must understand that this is not a minor hearing defect but deafness. We'll prescribe hearing aids for him, and we'll make an appointment with someone from the family support program." We were both stunned, shattered, in tears. What did this mean? Would he never learn to speak? What did we have to do? How must we bring him up? What must we do for him? In those first few days, we could do little more than struggle with our emotions: sorrow, confusion, shattered dreams, self-pity, guilt. Why? Why us? What caused the deafness? What exactly was wrong? Could anything be done to correct the problem? We needed help and advice, and we needed information. We had never even met a deaf person, as far as we could remember, and had little idea of what deafness was or would mean for us.

Our responses were the usual ones: guilt, mutual accusations, the search for a more hopeful second opinion, the need to understand "why"—if not in the sense of "why me," then at least in the sense of "through what cause." These are common reactions among parents of children newly diagnosed as deaf: the search for alternative opinions, for knowledge of what deafness is and what causes it, knowledge both abstract and personal. Like many others, I began avidly to read in the medical literature. From books on audiology and diseases of the ear, I soon learned about the parts of the ear: the outer ear, middle ear, and inner ear and their functions in the process of hearing. I learned that audiologists make a distinction between two types of hearing impairment. Where there is a problem with the outer or middle ear, they speak of "conductive impairment." That means that the problem lies not in the perception of sound but with its conduction to the places in the inner ear and brain where sound is analyzed. The most common cause is an infection of the middle ear known as otitis media. Sometimes it is only the temporary effect of a cold, and it can often be treated with drugs or surgically. That is quite different from "perceptive" or "sensorineural" hearing impairment. Here the problem lies in the inner ear, or somewhere between the ear and the sound processing regions of the brain. This is what most babies born with a hearing impairment have. It typically means that sounds of lower frequencies can be heard reasonably well, but that at higher frequencies nothing at all is heard. A child with this kind of a hearing loss would be unable to learn spoken language in the normal way. Nor can sensorineural hearing loss be treated, either medically or surgically.

Knowledge of the etiology may help with some of the processes of psychological adaptation, but it does not help with day-to-day coping. Certainly there were explanations we needed to eliminate in order to cope: a sense of personal responsibility or a sense of being punished in some inexplicable manner. We would have to think anew about the problem of communicating with Jascha and about his future education. Even with his new hearing aids, it was unlikely that normal speech would provide sufficient basis for communication or that speech would provide him with the cognitive and linguistic inputs he needed.

The NSDSK ran special communications courses for parents like us, and both the thrust of those programs and the other parents we got to know through them were profoundly to shape our attempts to cope with Jascha's deafness.

We were to be taught signs, the idea being that we would use both signs and speech. Using a combination of speech, signs, and finger spelling to discover the mix that works best for a given child was an approach that had grown rapidly in the 1980s. Educators speak of "total communication," a practice based on the idea that the effectiveness of communication is the most important thing. What medium is used matters less. We wholeheartedly agreed. We wanted to develop a way of communicating with our son as fast as possible: about his needs and wants, our demands, and almost as important, helping him learn about the world. We were given a list of twenty or so of the kinds of words we might want to use with a small child: simple words like cat, dog, eat, tree, drink. Then we were shown the signs that correspond to them. We set off home practicing with each other: cat, dog, eat, tree, drink.

The other parents we met at that time were a great source of comfort, in that they had gone through experiences like ours. We discovered how great is the human need to compare experiences. Many of those parents had children a little older than Jascha, and through their eyes we could see further down the road we would have to travel. Talking with other parents meant more than gaining information. It also offered an opportunity for telling our own story. Recounting what we had gone through to sympathetic and knowledgeable listeners was a crucial means of attesting to the validity of our experience.

The medical library, which had proved so rich a source of information on diagnosis and measurement of deafness, had little to offer on the more human aspects. What did the future have to offer a growing deaf child? What kind of an education might he expect to enjoy? What problems would he confront? Books by other parents of deaf children proved far more valuable than anything I found in the university libraries.

Pauline Shaw, the mother of a deaf daughter in England, wrote a powerful memoir of her experiences raising Judith.[2] As Shaw and her husband returned home with their eight-month-old daughter from a visit to the specialist, the tears flowed and "the long uphill struggle commenced." At age eleven, Judith was sent away to the Mary Hare Grammar School, a residential school in England that provides education to university entrance level for deaf pupils. At eighteen, Judith succeeded in gaining admission to university. I felt encouraged that a real education might be possible for Jascha as well. But again and again I came back to the early years of Pauline Shaw's story. How had they achieved all this? What would it require of Jascha, and of us, his parents? The NSDSK, on which we had become so dependent, seemed to look at things differently from the Shaws. Pauline Shaw devoted only a few lines of her book to communication using sign, recommending it only for children with multiple handicaps or with deaf parents. Her and Judith's success had been achieved through the endless practice of speech. We were being taught signs on the assumption that effective communication depends on their use. Who was right? Was there one best way? What to do?

In *Sing a Song of Silence*, published in 1983, Jessica Rees, a young woman of twenty, told of her own experience of deafness.[3] She had lost most of her hearing at the age of four, as a result of meningitis, and had spent her school years moving back and forth between schools for the deaf (including a period at Mary Hare Grammar School) and regular schools. At seventeen, Rees was the victim of a hit-and-run accident, and she lost the little hearing she had had. Rees's grandfather was headmaster of Eton College, one of England's most exclusive fee-paying schools, and her father a housemaster at the school. Her family, hardly surprisingly, took her education very seriously. Their efforts, the enormous effort put into teaching her to speak well, Rees's own determination to succeed in the hearing world, paid off. She succeeded in gaining admission to Oxford University. Like Judith Shaw, Jessica Rees was educated orally and her schooling offered little opportunity for using sign language. But while agreeing that good speech had been crucial to her success, Rees did not believe that sign language had to be avoided: "I usually speak normally while talking to hearing people or deaf people who lip-read and look down on sign language, or who do not know how to sign. But if I find myself in the company of deaf people who adamantly refuse to lip-read the entire conversation I am more than happy to converse in sign language and I refuse to feel that 'signing is always totally bad.' The important thing is to communicate, no matter how."[4]

The message of Rees's story was that good spoken language is essential to the deaf child while sign language can be an added advantage. She felt herself fortunate in her command of both. And yet it was hard not to feel that this was a clever and determined young woman still in the process of finding herself: "So I suppose you could say that I have the best of both worlds. But it is only recently that I have begun to think along these lines. Until round about a few months ago I felt pulled in both directions, unable to decide which was best and unsure as to where I actually belonged."[5]

Perhaps these uncertainties are not so surprising in a person of only twenty, deaf or not. David Wright, looking back over more than forty years of his life, had no such doubts.[6] Born in South Africa and deafened at the age of seven, Wright was brought up in Britain. His too was a remarkable success story, as well as the literary achievement one might expect from a man who had made a career as a poet and writer. Wright too had attended Oxford University. Are a university education and a literary career really a possibility for a deaf child? We could hardly believe it. Deafness seemed to be a minor thing in Wright's life and in his work. A poem reflecting on the significance of his fiftieth birthday does not even mention it.[7]

What to make of this? Could deafness be so insignificant? David Wright devoted some pages of the introduction to his book *Deafness* to a diatribe against another book on deafness, specifically on the history of deaf education: Harlan Lane's *When the Mind Hears*.[8] When first reading Wright's book, and still in search of any and everything that could help, I noted down the title as something to read later. Looking back at the introduction to *Deafness* I was later struck by how David Wright discussed what he saw as "not a history of the deaf or of deaf education" but a piece of "often tendentious propaganda for sign language" marked by "militant hysteria." Harlan Lane's thesis, that "oral instruction is a form of oppression of the deaf," was blatant nonsense so far as Wright was concerned.[9] Little did we realize at that time that what seemed to us a personal dilemma was also a battle of ideologies being fought out in article after article, in book after book. For us, the matter was personal and pressing. So much was at stake and there seemed to be so little time. Time was lacking not only because of our busy lives but far more importantly because our son was growing up fast. We had to make the right choices because perhaps there was no way back. Who to believe? On what to pin our hopes? Should we make every effort to learn to sign, through our fortnightly lessons and the videotapes that the NSDSK gave us for home study? Should we devote ourselves to auditory training like Pauline Shaw had done: one and a half hours every day?

William McKellin, a Canadian anthropologist who also had a hearing-impaired child, wrote about families like ours.[10] One of the most difficult things is the conflicting advice. There are abstract analyses written by each side: by the oralists stressing the use of speech and by the manualists emphasizing the use of signs, and each had given rise to what McKellin called its own "folklore." Each had its powerful personal narratives. I had already come across two examples of oralists, and Pauline Shaw's and David Wright's books made a great impression on me. Jessica Rees's message was less clear-cut. It was to take longer before I came across no less powerful and moving a book that offered precisely the opposite message to what Shaw and Wright had to say.

James Spradley, an anthropology professor, and his brother Thomas started writing *Deaf Like Me* when Thomas's daughter Lynn was seven.[11] Lynn had been born deaf, a consequence of her mother's German measles. This book too described the shock, the guilt, and the desperation of a family confronting deafness for the first time. This family too sought desperately for information, read avidly on deafness, subscribed to the *Volta Review*,[12] wrote off for the John

Tracy Clinic correspondence course,[13] worked as determinedly as Pauline Shaw had done at teaching their daughter to speak. But somehow things did not work out as they had done in all the success literature they had been devouring. Terrible tantrums followed as their daughter, passing her third and then her fourth birthdays, still could not communicate her needs and wishes. Lynn's parents, saddened and frustrated, persevered as best they could. Perhaps they had to work harder at it. By the age of five, Lynn had mastered only a handful of words despite attending a good oral school for deaf children with a well-qualified staff. Statistics suggested that many deaf children never learned to speak comprehensibly. What if their child was one of those? A small group of parents, dissatisfied, raised the possibility of introducing a sign language class at the school. A bitter debate broke out. The Spradleys got to know some of the dissatisfied parents, visited them at their home, saw their communication with their little girl. They could not believe what they saw. Unsure, worried at frictions that might arise with the school, they nevertheless decided to learn a few signs and try them out. The effects were instant and magical: "Thirteen words in less than a week. Almost three times as many as she had learned to speak in five years. We were totally unprepared for the speed with which Lynn now began to use signs spontaneously."

As the authors, Lynn's father and uncle, reflect on the changes that were so rapidly affecting their family, the tone becomes bitter. How could they have been advised, obliged even, to place the total burden of responsibility for effective communication on a small child? How could they have been forced to forbid her to use her "native language," sign language?

Two very different kinds of narrative, each of which spoke powerfully to the experiences we had had. William McKellin pointed out that a family's choice, for example for one or another mode of communication, depended very much on that family's particular circumstances. Many parents have contact principally, or even exclusively, with professionals who stress the use of spoken language. Allowing a child to communicate by means of sign, from their point of view, is impeding the child's acquisition of what they refer to simply as "language." That was certainly the case in some of the books I had read. How to deal with the convictions of the therapists and teachers on whom, as a parent, one is so dependent? It took a willingness to face conflict for the Spradleys to start communicating with Lynn using signs.

Most of the other parents we were getting to know were in roughly similar positions, and many faced similar doubts and uncertainties. Everyone felt the need for more information. A previously unknown sign, some piece of news about the deaf world, something happening abroad: all of these were offered and received as gifts to the group. The importance of what we shared produced an intense feeling of solidarity in a group of people who would probably otherwise never have met. Part of it was through providing each other with the "witnessing" of the stories we wanted to tell. Part of it was simply giving each other advice and information. In the short term, we had to press ahead with learning sign. Just as we'd been advised, we tried to communicate with Jascha through spoken Dutch accompanied by the signs we were learning. English, my mother tongue, had no place in a scheme that was demanding enough for us and infinitely more demanding for Jascha. But still, we needed to know more, and the hope of a cure had not wholly vanished.

I no longer know when exactly we first heard of the cochlear implant, although in March 1989 I had clipped an article about it from a newspaper. In any event, from this article I must have had some sense of a possible miracle. Under the headline "Deaf Woman Amazed to Hear Sounds Again after Implantation," the article reported on how a woman deafened forty-three years earlier as a result of meningitis had been one of the first people in the Netherlands to receive a new twenty-two-channel Australian implant.[14] Six weeks after the operation, with the device working well, she could hardly believe what had happened. Professor van den Broek, the surgeon who carried out the operation, was less astonished: "We have to be very careful in arousing expectations, but I think that people who are 100% deaf really can be helped."

Clinging to the hope that medical science could do something for Jascha's deafness, I could hardly help feeling that this was it. This astonishing new device began to loom large, with intellectual curiosity and personal hopes reinforcing one another. We felt we had to pursue every avenue, and in those first months the cochlear implant seemed the best hope. From what we had heard or read, the device seemed to enable hearing for at least some deaf people, even those suffering from total sensorineural deafness.We didn't know much of what it entailed, but four or five months after Jascha's diagnosis we certainly wanted to find out if this device could set our lives back on the course we'd imagined. At the end of 1989, we visited a professor of ear surgery at a nearby university. As Anja described the visit, "So far as cochlear implantation was concerned, [the] professor explained to us that, perhaps, in the future, it could have real therapeutic benefits to offer. 'But if I was in your position, I wouldn't bet on that. Not now.' Professor Smit[15] was a man whose manner and whose explanations inspired trust. We had the feeling that his way of looking at things fitted us: fitted our way of thinking and our way of dealing with things. It gave some peace."

Professor Smit inspired confidence in both of us. We decided not to proceed with a cochlear implant for our son, at least not then. But I was intrigued by my own initial expectations of this apparently miracle technology. Where had it come from? How could it arouse such hopes? Was sign language really an alternative? We had decided to follow Professor Smit's advice, but was it wise to put one's trust in a single expert? Increasingly intrigued by the technology and by questions like these, the idea of a study of the cochlear implant began to grow on me. What kind of a study should it be? The history of the device—who had developed it, where, why—would probably be the most straightforward part. But since I also wanted to tie in my questions about hope, alternatives, and trust, I'd have to think carefully about how best to proceed. The starting point, in any event, would be the existing historical and social science literatures dealing with new medical technologies. It is quite a substantial literature, and the first thing was to see what I could take from it that would help me think through my own study. I decided to structure my research according to the questions that grew out of those first reflections on the implant. Where do new medical technologies come from and how do they arouse such hopes? What does it mean to speak of the possibility of alternatives? And how do we, as individuals but also as societies, decide whether a new medical technology is a good thing?

Sources of Technology, Sources of Hope

Faith in the possibilities of medical science is nothing new. For more than a century the widespread expectation, at least in the industrialized world, has been that the population would become ever healthier thanks to science and to medicine. Most of us share this faith in medicine's ability to protect or cure us. Our children no longer run the risk of being crippled by polio. We appreciate all the efforts to reduce the risks we run and eliminate obstacles to our living long and healthy lives. Since there are now therapies to slow down or even reverse processes of aging, we need no longer accept the decline of our bodies as the years go by.[16] Vaccines to stop us from getting fat or become addicted to nicotine and drugs to keep us in top condition without the need of exercise are not far off. Researchers are even working on brain chips to give us knowledge without the need to study. Since we believe a solution is just around the corner, we are susceptible to media announcements of a "miracle cure."

This faith seems justified. In little over a century, scientists have discovered and invented an astonishingly long list of life-saving and life-improving drugs and devices. Producing them became the business of two distinctive industries. The pharmaceutical industry was born out of the nineteenth-century chemical industry as chemical companies began to create synthetic compounds for medical use. Analgesics and antipyretics such as aspirin, diphtheria and tetanus antitoxins, Salvarsan for treating syphilis: all emerged around the turn of the twentieth century.[17] In the 1920s and 1930s, the pharmaceutical industry developed vitamins, hormones, insulin, and the sulfa drugs, while penicillin came soon afterwards. Quite separately, and largely inspired by Willem Roentgen's discovery of X-rays in 1896, manufacturers of electrical equipment such as General Electric and Philips laid the foundations of a new medical devices industry.[18] Primitive early X-ray imaging devices were improved, increasingly in directions determined by the requirements of the new specialty of radiology. Specialized devices were developed corresponding to the distinctive needs of imaging different parts of the body. Later, ways of using the power of the computer to extract more information from X-ray images were found, and the CT scanner became a major addition to the armamentarium of radiology. Quite different ways of generating images, using ultrasonic waves and the subtle properties of complex molecules in the body, added still further to the tools available.[19]

Innovation processes in these two areas, drugs and devices, tend to differ. Whereas the development of a new drug generally begins in the laboratory of a pharmaceutical company, development of a new medical device might well begin in someone's garage, with some kind of "tinkering." A physicist or engineer has an idea that some new principle or material could be applied to a medical problem. Though not in his garage, that was how Godfrey Hounsfield came to develop the CT scanner in the 1970s. Or, a technically inclined physician or surgeon, dissatisfied with an existing technique, begins to sketch out what a better one might be. This was how Willem ("Pim") Kolff conceived of the artificial kidney in Nazi-occupied Holland and Scottish obstetrician Ian Donald the technique of obstetric ultrasound, or echoscopy. As initial tinkering begins to look promising, new skills and competences—medical, technical, and industrial—are drawn into the project. The beginnings of both the artificial hip and the intraocular lens (used to restore vision to people suffering from cataracts) were like this. Like the Echoscope, both were initially conceived by British surgeons just after World War II.[20] In each case a surgeon-inventor (John Charnley, Harold Ridley) sought assistance regarding materials and manufacture from a local craft-based company.

Then similar collaborations may occur elsewhere. Typically these different groups of scientists and engineers make different design decisions: the details of the technology are not fixed in advance but emerge through interaction with potential users. Emphasizing the importance of these interactions for the form that a technology ultimately takes, sociologists speak of a technology being "socially constructed." Sometimes disputes regarding the readiness of a new medical device for use with patients arise. The work of development then proceeds in a network of collaborative and competitive relationships involving hospitals, manufacturers, and laboratories: what economists call a "distributed innovation system." Competition for professional and scientific esteem between clinical investigators gives rise to commercial competition as devices enter production. As technologies mature, so the shape, the balance, and the locations of the innovative networks shift. Large companies come to dominate production, and further development of the technology then takes place in their laboratories.

The main work of innovation is shared among scientists, engineers, physicians, and surgeons, working in hospitals, universities, and the business sector. How committed they are to innovation and how much effort they put into it depends on what they expect the rewards to be and their assessment of the chances of success. Since pharmaceutical and medical device innovation have for decades been seen as valuable to the national economy, a wide range of government policies and initiatives have been deployed to stimulate them. Subsidies or tax incentives for R&D can reduce the risk involved for private investors or facilitate expensive research. Or public funds might be used to guarantee markets and so encourage companies to address areas where market prospects might look unattractive but where there is a great social need. Sometimes, as in the case of the CT scanner, a government department plays an important "broker" role in bringing partners together. The broader context in which innovation takes place, including all kinds of political commitments and government programs, influences how and for what purposes new medical technology is developed. Healing the scars of war was a major stimulus to development of new medical technologies throughout the twentieth century. World War I, among its many horrors, led to thousands of soldiers returning home without an arm or a leg. They were maimed, but they were also heroes disabled in the service of their country. There was a widespread feeling that everything possible must be done to enable them to reintegrate into the community as best they could. Providing them with good artificial limbs was to be part of this collective provision, a commitment out of which was born the modern prosthetics industry.[21] World War II led to a host of new technologies as new uses were sought for expertise demobilized as the war ended. Diagnostic ultrasound came about after redeploying skills, materials, and technologies developed in the war for medical purposes.[22] Donald, Charnley, and Ridley were all influenced by their wartime experiences.

They were also influenced by the massive reorganization and nationalization of health care introduced in Britain after the war. Government policies or legislation developed with quite other objectives can have important indirect influences on the development of new health care technologies. For example, the Americans with Disabilities Act (ADA), passed in 1990, prohibits discrimination against individuals with disabilities. It also requires that institutions ensure their premises and services are accessible to people with disabilities. The adaptations to comply with the law stimulated R&D and design work in assistive technology.[23]

Today an integrated medical device industry increasingly resembling the pharmaceutical industry seems to be emerging: one dominated by a small number of multinational corporations competing in and for global markets.[24] Globalization has had both direct and indirect influences on medical innovation. For the pharmaceutical industry in particular, developing countries have become an important locus for the establishment of clinical trials, and large developing countries have become prized markets. The influence of this industry in setting the rules of global trade and in global public health forums as well as its tendency to price new drugs beyond the reach of patients in poor countries have received a good deal of critical attention. Valid though criticism has been, recent anthropological research has shown that these companies, in alliance with civil society organizations, sometimes play an important part in pushing access to essential drugs to the top of the political agenda. Out of this movement come both a "pharmaceuticalization of public health" and a stimulus to domestic pharmaceutical industries in countries like Brazil and India.[25] Gradually, these local industries are moving beyond the production of generic drugs to become a new and potentially distinctive source of drug innovation.

Beyond market liberalization and globalization, in all its many senses, other contextual changes have affected the way in which new medical technologies are developed. Of course, vaguely taken to stand for a heterogeneous assemblage of factors—economic organization, social policies, values, and so on—the context of technological change is always in process of evolution. To be more specific, the change I want to emphasize here concerns the relationship of users to technology, a relationship that now attracts a lot of attention from scholars working in the field of science, technology, and society (STS).[26] Central is the relationship of the end users of medical technologies—we can refer to them as "patients" though that is not always how they refer to themselves—to the development and provision of new health technologies. Chronically sick and disabled people in the wealthy industrialized world—or at least an articulate elite among them—have become more demanding, less willing to accept restrictions on life style than they were. Especially young people with disabilities, or their parents, are no longer willing to put up with drastically diminished lives compared to their peers. People suffering from chronic illnesses have also become more knowledgeable. Through the Internet it has become far easier to learn directly from other patients, to share experiences in discussion groups or chat rooms, and so draw on a source of knowledge potentially different from that of the medical profession. We hear a great deal these days about "patient empowerment."

In Britain, with its state-run National Health Service, encouraging and enabling patients to become "informed consumers" of health care is a major component of government policy. What this usually means is what it explicitly says: making an informed choice among available items for consumption. For the producers of health technologies, drugs in particular, this consumerist orientation has been an opportunity to be exploited to the full. Where, as in the United States, direct-to-consumer advertising of drugs is permitted (it is forbidden in Europe), the mass media have been used to heighten awareness and stimulate demand. A study of the way in which the drug Tamoxifen was advertised to healthy American women illustrates how subtly this is done.[27] Tamoxifen is a breast cancer chemotherapy drug that, at the time of the study, was being advertised for preventive purposes, as a means of reducing the risk of getting breast cancer. Advertisements appealed to notions of empowerment and consumer competence, urging women to assess their personal risk ("if you care about breast cancer, care more about being a 1.7 than a 36B") and take preventive action.[28] Use of the drug for preventive purposes would vastly increase the market and clearly offered major commercial benefits to its manufacturer.

The "expert patient" imagined by policy makers and by advertisers is one willing and able to manage his or her own illness, "in partnership with" the health care provider. The "experience-based knowledge" exchanged and consolidated in Internet-based discussion groups then assumes a medical understanding of what is wrong and the willing consumption of medical services and technologies. British sociologists studied an Internet forum devoted to obesity and particularly to the use of a weight-loss drug, Xenical. They found postings that covered not only technical matters surrounding use of the drug but also the wider experience of being overweight. And while many criticisms of the medical profession were expressed, participants "accepted that overweight must be treated, with Xenical as a method of weight reduction."[29]

Sometimes, however, users are dissatisfied with the services or the technologies they are offered. In designing medical, as any other, technologies, all kinds of assumptions regarding the intended users—their heights and weights, competences, preferences, behavior and values—are made. But the user "inscribed" in a technology, imagined by its designers, may not correspond with real users in the real world.[30] Their needs and preferences may differ from what the designers had in mind, or technological change may have failed to keep up with changes in their needs or aspirations. The consequent mismatch can evoke a variety of responses.[31] For example, recipients of the artificial hip in the 1960s and 1970s were mostly elderly people with low levels of physical activity. But by the 1990s, there were also many younger candidates who wanted the possibility of long-distance walking, swimming, dancing, or playing golf—activities their predecessors had not dreamt of. Older designs were no longer adequate to their needs (or demands), and new ones had to be developed.[32]

There are now numerous instances of people with disabilities, frustrated at the inadequacies of existing technologies, themselves developing devices that better meet their needs. A deaf physicist, Robert Weitbrecht, invented the text telephone (TTY) in the 1960s because he believed that deaf people like himself had a right to the convenience of telecommunications, and he was able to see how that might be possible.[33] The story has been told of how, in 1978, a young Californian woman named Marilyn Hamilton crashed her hang glider into a mountain, sustained a spinal cord injury, and became a paraplegic. Hating the cumbersome hospital-issue wheelchair she was provided with, Hamilton got a couple of her friends, hang glider pilots like herself but also inventors, to develop a sporty lightweight model for her. Aware of how much of an improvement their new design was over existing technology, the three of them went into the wheelchair manufacturing business, offering their stylish product in a variety of screaming neon colors. The new design was successful: "Hamilton's proud chairs struck a chord with the emerging disability rights movement," then growing in numbers and in its demands for access and independence.[34] In a Ugandan border village, physically handicapped people displayed their resourcefulness in converting bicycles into devices with which they could earn a livelihood transporting goods across the Kenyan border.[35] Weitbrecht, Hamilton, and the Ugandan villagers all required technical skills as well as social and economic capital to conceive, realize, and perhaps market innovations adapted to the needs they identified. But the skills and the capital required differed from one context to the other.

Many users of health technologies, lacking these skills or resources but no less discontented, make personal adjustments of a more private kind. Even under the surveillance of a medical practitioner, users of medical technologies sometimes make adjustments that depart from the instructions they were given. They try to bring the device or their drug regime into better alignment with their reading of their own bodies, with how they want to live, or with the image they want to project. For example, it has been known for decades that some 50 percent of patients do not take their medications precisely as prescribed: so-called noncompliance. There is a kind of "experimentation" involved, as people try to establish what a particular medication means for them in terms of the acceptability of side-effects, the minimal dose that seems to work for them, and possibly by incorporating its use in a broader coping strategy involving complementary therapies or lifestyle changes.[36] Rejection of professional advice may go beyond experimenting with a drug, perhaps as far as refusing a test, drug, operation, or vaccine. Some pregnant women reject prenatal diagnosis. Their rejection turns out to be a complex and varied matter. Some highly educated women reject testing on the basis of philosophical, religious, or ethical reasoning, on the basis of their own prior experience, or because they want to make their own assessments of the risks they run in their pregnancies. Having studied the medical literature, their explanations are inflected by biomedical discourse. Less-educated women may assess risk by reference to the pregnancies of friends and family living similar lives, reasoning on the basis of "a practical sense of community epidemiology."[37] Some African American couples refuse because they mistrust medical science. They may have heard of the many ways in which, in the past, black communities have been exploited for research purposes. For them, individual choice is contextualized by "racially differentiated sentiments concerning medical intervention and experimentation."[38]

In recent decades, and aided by the networking possibilities provided by modern communication technologies, patients in North America and Western Europe (though much less so elsewhere) have joined together. These patient organizations, or organizations of parents of sick or disabled children, offer members opportunities for the exchange of information and the "witnessing" of experiences. Beyond that, patient organizations often demand improved facilities or better access to treatments for people suffering from the particular disease or set of symptoms around which they are organized.[39] Advocacy sometimes goes further. Sometimes advocates try to gain influence over the research and development agenda, over the "translation" of laboratory research into viable therapies, or over the processes through which new drugs or treatments are licensed.[40] Intrigued by the question of how such influence is achieved, by the possibilities of democratizing the research process, STS scholars have studied how and by how far they have been able to do this. Mary Anglin showed how breast cancer activists in the United States gained representation in research funding committees and influenced allocations of funds to breast cancer research.[41] Steven Epstein showed how AIDS activists (again in the United States) were able to influence HIV research and speed up the introduction of new therapies. Vololona Rabeharisoa and Michel Callon classified the different ways in which patient associations are engaged with research. They analyzed the strategy followed by a French association concerned with neuromuscular diseases, the Association Française contre les Myopathies, in gaining recognition for muscular dystrophy as a disease that could potentially be controlled and in influencing the research process.[42] In each of these fields, the focus was on the search for drugs that would save lives. Activists whose lives were at stake, or whose children's lives were at stake, were able to force things along faster or in different directions.

In order to do this, they had first to convert their personal experience into cultural capital. Personal experience has no inherent authority and is not by definition a source of expertise. The extent and the ways in which it is acknowledged as relevant differ from one society to another and from one sphere of action to another. In some parts of the world, the experience of living for years with a chronic ailment is now accepted as leading to a certain sort of knowledge. Patients acquire knowledge based on experience of the way their own bodies respond to treatments.[43] In today's health discourse, they are acknowledged as "expert" in their own illness. Doctors who have treated such patients over long periods of time are said to respect their insights and their judgments and to work with them in determining, for example, the optimal dose of a medication. This is the kind of expert, empowered patient that we are encouraged to become—one who not only understands his or her own illness but also takes responsibility, in partnership with the physician, for its management. The unstated assumption is that the patient's experience has been sufficiently refracted by, or imbued with, medical thinking, that his or her management strategy will not depart too radically from professional advice. In a study of an obesity Internet discussion group, it was found that most participants were looking to develop precisely this kind of expertise: "'Official' advice from doctors and other was refracted through the lens of experience, but there remained an underlying commitment to medical models of illness and treatment."[44] A patient setting off on a different road, calling medical models or explanations into question, will not be granted the same tokens of acknowledgment. For example, a mother rejecting vaccination for her child is likely to encounter anything but respect for the reasoning behind her decision. This is all the more true when what is at stake is not self-management but how in general a particular condition or patient population should be treated. As we move from the private to the public sphere, as organized patient groups publicly question the ways in which a particular condition is treated, the status of experiential knowledge is opened to contestation.

Matters are rendered more complex by the fact that the experience of having lived with a particular illness or disability can be narrated in very different ways, as the various accounts of deafness suggested. Patients develop different accounts and competing sets of demands based on them. An ethnography of breast cancer activism in the San Francisco Bay Area shows this particularly clearly. In an insightful ethnographic analysis, sociologist Maren Klawiter contrasted the annual get-togethers of three activist groups at each of which women who were living with (or had had) breast cancer told their tales and (in some cases) exhibited their bodies.[45] But the objectives of the three groups were radically different. One group, clearly sponsored by the pharmaceutical and medical care industries, aimed at raising money for cancer research. Another group, less committed to biomedicine, supported self-help and less aggressive therapy. The third group had a very different message: that the industry marketing anticancer drugs is the same industry that, through production of pesticides and insecticides, is also contributing to high rates of cancer. No corporate sponsorship was to be seen here, and this "Cancer Industry Tour" began with a protest outside Chevron's corporate headquarters. Demonstrators chanted, "Stop Cancer Where it Starts. Stop Corporate Pollution!" and "Toxins Outside! Cancer Inside! Industry Profits! People Suffer!"

Two competing trends are now at play, each of which bases its claim to legitimacy on an appeal to the notion of patient empowerment. One rests its claim on the "informed consumer," able and willing to help determine his or her own treatment and claiming the right to do so. The other, espoused by the more radical leaderships of some health advocacy organizations, may deny the appropriateness of the treatments on offer and may claim that patients are misled by commercial interests. Conflicts about the appropriateness of tests or drugs can emerge between radical leaderships and more conservative rank and file. Again, it is difficult to generalize because whether or how this occurs seems to depend on the way in which health care is organized, and in particular on the role of commercial interests.[46] Linda Hogle's study of the marketing of the anti-cancer drug Tamoxifen found that women surveyed had confidence in their own ability to judge the claims made in advertisements. By contrast, breast cancer activists felt that advertising was misleading and expressed concern that "physicians may just give the drug to women who request it, bowing to pressure from consumers, and expressed considerable concern about women's ability to interpret information presented in the ads."[47] A controversy over this advertising campaign arose as cancer activists mobilized against it, placing their own full-page ad in the *New York Times*. Like one of the groups that Klawiter studied, this coalition (Breast Cancer Action) referred to the drug as a carcinogen and warned women of the commercial interests behind the information they were being given. Many of the women Hogle interviewed objected to what the activist coalition was saying. Whatever the accumulated expertise of the cancer activists and whatever the commercial interests behind the drug advertisements, the women interviewed responded more positively to the drug ad than to the activists' warning.

That the mass media played an important role in this dispute should come as no surprise. Newspapers and television are far more focused on health today than they were a few decades ago, and few developments in medical science escape their attention. As they report new developments, the message they nearly always try to convey is inspired by the future possibilities of medical science: a message of hope.

In the years after World War II, some began to imagine a world in which miniaturized electronic substitutes would serve as long-term replacements for organs that had worn out through age or had become damaged through injury. In 1955, these visionaries established the American Society for Artificial Internal Organs. Willem Kolff, then working at the Cleveland Clinic, was elected chair of the new Society.[48] By 1965, a *New York Times* science writer could look back and see the dream becoming reality: "Within approximately the past decade have come a remarkable series of developments: artificial tubes to replace segments of human arteries, artificial heart valves, artificial heart pace making devices. Even more ambitious aids, or substitutes, for vital human organs and systems are being perfected in laboratories."[49] Each new discovery was accompanied by "excitement, publicity, and renewed hope."[50]

Examples could be listed indefinitely. For example, much has been written over the past few years about the antiretroviral drugs that have turned HIV from a killer into a chronic disease, at least for those with access to them. Harvard physician-anthropologist Paul Farmer has described how these drugs "captured the imagination of providers, pharmaceutical companies, and, especially, patients" when they were announced at the Eleventh World Conference on AIDS in July 1996.[51] It was not long before *Newsweek* asked on its cover "The End of AIDS?" A few years ago *Scientific American* published an issue with the title "Your Bionic Future."[52] It had sections on "your new body," "your new mind," "your new senses." Pharmacogenomics should enable clinicians to select people who will respond to a particular drug or therapy. The therapeutic use of embryonic stem cells should make possible the (re)generation of damaged tissue. Biochips could be used not only to correct deficient senses (hearing, vision, smell, touch) but to enhance the sensory capabilities of people we now regard as normal. "Your bionic future" suggests that through new uses of genetics, cellular biology, and neural implants, the human race could enjoy much longer lives in enhanced and beautiful bodies.

Medical technologies have become much more heterogeneous. The once clear distinction between drugs and devices has no place for the new hybrids, like artificially engineered tissue, that seem to be both or neither.[53] At the same time, the processes through which they emerge involve a more and more complex array of increasingly heterogeneous participants. The initial collaboration between a physician or surgeon and an industrial technologist, in which a new device might be conceived, is now shaped and influenced by (national) economic interests, by professional interests (that may conflict with other competing professional interests), by government policies towards the organization and finance of health care, and by the hopes and expectations of large numbers of individual patients. The responses of patient advocacy groups, where they exist, may also have an important influence, though it can cut both ways. Sometimes, as in the area of HIV/AIDS or neuromuscular diseases, patient advocacy has supported the search for more effective treatments through raising funds or political lobbying. Sometimes, however, activists have urged caution, fearful that unrealistic hopes were being encouraged by commercial interests. A cautious, or skeptical, perspective—one critical of industry, of industry-sponsored science, or of the hopes held out—is unlikely to receive favorable media coverage. The mass media prefer to convey a more optimistic message, and this is the one most of us—patients, parents, friends, and relatives of patients—will pick up.

This was the context in which I had to try to understand where the cochlear implant came from. I had to be sensitive to the perspectives of the various participants, to the interplay of interests through which it took shape, and to the role of advocacy. I also had to pay attention to role the mass media played in encouraging people—not least parents like me—to expect so much from the device.

Can There Be Alternatives?

Despite the optimistic faith in technological progress that has marked western culture for more than a century, skepticism in the face of medicine's reliance (or overreliance) on technology is also not new.

The atomic bombs dropped on the Japanese cities of Hiroshima and Nagasaki at the end of World War II provided powerful evidence that the marriage of science and technology did not lead solely to material progress. Gradually, through the 1960s, it was beginning to seem that science itself had become distorted by military and industrial interests, and instances of the destructive power of science-based technology were becoming a source of widespread concern.[54] Finding themselves in need of new justifications for the investments in increasingly expensive instruments that their research required, scientists turned to history. Listing the many benefits that science has provided in the past, they suggested that the future would yield a similar harvest: a problematic extrapolation that has been referred to as "the myth of infinite benefit."[55] In the field of medicine, doubts emerged too, but it took longer. For a few critics, scientific medicine was becoming not only the respectable face of an evil capitalism but it was actually dangerous for health.[56] This radical intellectual critique acquired some unexpected allies. An eminent British professor of social medicine, Thomas McKeown, argued that the decline of infectious disease had not been due principally to the development of drugs and vaccines but to improvements in living and working conditions.[57] But perhaps the most influential critics of technological medicine were the economists. As the costs of health care rose inexorably, governments began to worry. The cost of a day in the hospital in the United States rose from fourteen dollars in 1950 to more than one hundred and fifty dollars by 1976, and new technology seemed to be responsible for much of the rise.[58] Health economists, a new profession in the 1970s, convinced governments of the need not to be swayed by medical optimism but to look critically at the costs and benefits of each innovation as it emerged.

At the same time, consumers of health care were also starting to ask difficult questions. This was particularly apparent in relation to the technologies of reproduction and birth that became a focus of attention for the women's movement. Hundreds of thousands of women had already learned, to their cost, that the benefits of advances in medical technology could not be taken for granted. The drug thalidomide, taken by many pregnant women as a sedative in the late 1950s, had resulted in the birth of thousands of children suffering from deformities of the limbs. Another drug, diethylstilbestrol (DES), had been prescribed to millions of pregnant women over some decades for the prevention of miscarriage. From the early 1970s onwards, it had become clear that daughters born to these women had significantly increased risk of reproductive tract abnormalities and a variety of cancers. In the early 1970s, millions of women started to use an intrauterine contraceptive device known as the Dalkon Shield. It was later established that the device could cause infections of the uterus that could later lead to miscarriage. Women surely had reason to be wary. Feminist social scientists working with the women's movement began to look critically at how medical technologies were developed, tested, and then used on women. Anthropologists, sociologists, and historians looked at technologies of prenatal diagnosis (including diagnostic ultrasound) very differently either from technological enthusiasts or cost-conscious economists. Their often highly critical analyses of how physicians used these technologies drew both on their analytic skills as social scientists and on their personal experiences as women.[59]

By the 1980s, the dream that had inspired the artificial organ pioneers seemed to some to be turning sour. After decades of studying transplantation medicine, sociologist Renée Fox and historian Judith Swazey decided the time had come to "leave the field." The "intensity and expansion of the drive to sustain life and 'rebuild people' through organ replacement has progressively alienated us, particularly the unquestioning and even celebratory way in which the transplanting and retransplanting of virtually every organ of the human body is creating larger and larger numbers of 'patchwork men and women' whose quality of life is dubious at best."[60] It was not only the biological reductionism involved in treating people as collections of organs (for later reuse), or genes, but the unquestioning and unflinching ardor with which therapeutic goals were being pursued, the inability to say "enough is enough," that Fox and Swazey found unbearable. Development of transplantation programs was not only diverting attention and resources from the health care needs of the majority of citizens, it was also exacting too high a social and an ethical price. Perhaps medical science had gone too far in doing all it could to provide the affluent few with everything they could desire ("to manage mood and affect, and to rescue people from the folly of their personal choices and hazardous lifestyles").[61] Individualization, extended into the domain of risk by developments in genetics and pharmacogenomics, fused with a growing fascination with bodily aesthetics and functioning. The consequence has been that "by the start of the 21st century, hopes, fears, decisions and life-routines shaped in terms of the risks and possibilities in corporeal and biological existence had come to supplant almost all others as organizing principles of a life of prudence, responsibility and choice."[62]

Over the course of three decades, the critique has been refined and extended. It has gradually made clear that what is at stake goes beyond the use of complex technology to exert and reinforce professional dominance, as suggested by early critics of "medicalization."[63] Theoretical analyses today focus on the ways in which modern society has become dependent on new life-science-based technology, on the irresistible promise of indefinite advance, and on the ways in which fundamental values (human dignity), concepts (identity, kinship), and dichotomies (life/death, human/non-human) are eroded or transformed. Theorists are less concerned with alternatives, with "how it might be otherwise." Yet the field of health care today also shows numerous very different attempts to articulate alternatives.

Patient organizations play an important role in this articulation of alternatives. Their formation and growth can be understood not just in terms of the possibilities offered by the Internet, but in terms of a growing tendency to identify with others sharing a common biological characteristic. The biology-based social groupings that have become increasingly common—some refer to "biosocialities"—have become a new force in contemporary politics.[64] Some scholars have pointed to demands that go far beyond the politics of health care provision, challenging the vectors that lead from biological difference to stigmatization and social exclusion. As sociologist Nikolas Rose puts it, "They use their individual and collective lives, the evidence of their own existence and their vital humanity, as antagonistic forces to any attempt to re-assemble strategies of negative eugenics within a new exclusionary biopolitics."[65]

In trying to assess these "antagonistic forces" or the ways in which patient organizations have tried to articulate alternatives, one must be careful. Looking at the world from the United States, Britain, or the Netherlands, it is easy to overestimate the extent to which sick or disabled people are able to mobilize and to advocate on their own behalf. A great deal stands in the way of such collective mobilization, and in much of the world these patient organizations scarcely exist. Where disability or chronic illness is highly stigmatized, as is still the case in many countries, few disabled or sick people are likely to identify in public with the source of their stigma. They are more likely to do everything they can to conceal their affliction.[66] The history of organizations of people with disabilities, where they exist, shows how difficult collective mobilization has been. Such histories also show how the form such organization takes has depended on traditions of social welfare, of collectivism as against individual rights, and of protest. The American disability movement was inspired by the civil rights campaign and by the women's movement of the late 1960s. The Vietnam War and the large number of young men who came home disabled was another stimulus to mobilization. In other countries, mobilization of people with disabilities, when it occurred, took different forms.[67]

In addition to variations between countries, disability activism has had different objectives than those of the HIV/AIDS activists studied by Steven Epstein or the parents of children with neuromuscular diseases studied by Michel Callon and Vololona Rabeharisoa. Rejecting notions of patienthood or a medical model, people with disabilities have mobilized around a range of issues restricting their capacity to participate fully in society: essentially rights-related issues.[68] In the United States, a surge in protest activity on the part of people with disabilities occurred in the late 1980s. One causative factor might have been the highly publicized protest by deaf students at Gallaudet University in Washington, DC, demanding the appointment of a deaf president for their institution.[69] That protest, it has been suggested, contributed to a change in political culture regarding disability and greater sympathy for and interest in disability issues in Congress than had been the case previously.[70] It was in that year (1988) that a first version of the Americans with Disabilities Act (ADA) was introduced in Congress. For the disability rights movement, it was not physical impairment as such but stereotyped attitudes and disabling practices that were at the root of the problem. From such a perspective, medical practice is of little significance, and the collective struggle is not for improved therapies or more biomedical research but for civil rights.

In studying cochlear implantation, I had to explore the relationship of deaf people to development of the implant. It was important to understand the nature of their organization and involvement: how far, if at all, they had advocated for or against, or otherwise influenced its development. I also had to think hard about how the use of sign language relates to the medical option and about the sense in which it was an alternative, possibly rooted in a rights-related discourse comparable with that of disability rights advocates.

Trying to Decide What to Do: Adjudication, Deliberation, Mediation

The doubts we had had about a cochlear implant for our son were similar to those many people have to deal with, trying to do their best for themselves or their children and faced with different treatment options. Despite our initial hope that the cochlear implant was the miracle we wanted, it was not the choice we ultimately made for Jascha. Professor Smit advised us not to pin too many hopes on it, and because we trusted him we followed his advice. Were we right to trust him? How could he be sure? What do medical specialists base their advice on? The stream of new treatment options is now such that not only patients and parents find it hard to know what is best, so too do medical professionals. Trying to do their best for their patients, they too need to find ways of judging the claims made for this or that new drug or new procedure. Nor do the doubts stop with doctors. Health policy makers and managers in the health insurance industry have a similar problem. They too have to deal with a continuous flow of new tests, vaccines, drugs and devices, for all of which great claims are made, but not all of which can be provided. Resources are limited and have to be used both judiciously and justifiably.

Like health policy makers and managers, the medical profession has long put its faith in scientific evidence. Thirty years ago, it became apparent that new medical technology brought with it costs as well as benefits, and that "promising" new interventions were being introduced on the flimsiest of evidence that they were in fact beneficial. As the costs of health care rose out of control, ways were sought of more critically evaluating new drugs and devices, of determining which benefited patients, and how much. Better and more comprehensive data, including data on costs and benefits, were the key to more rigorous and rational decision making. Gathering and synthesizing data related to each intervention became central to the field of medical technology assessment (MTA).

Introduction of MTA did not solve the problem of making decisions that would be accepted as both rational and unbiased. Within the health sciences, researchers were showing that much medical practice was based less on systematic evidence than it was on tradition or the attitudes of senior members of the profession. Hard-to-explain national differences were identified.[71] Not only that, but much of the evidence justifying treatments that was to be found in the medical literature was of poor quality. One response to this, from within medicine itself, was what is now known as "evidence-based medicine."[72] Key actors were mostly clinical epidemiologists, and their principal claim was that a more rigorous use of epidemiological methods and evidence would lead to better medical practice. Evidence was not all of the same value, or quality, and in this "hierarchy of evidence," the greatest degree of confidence should be accorded to the randomized control trial. Growing numbers of public health doctors, health researchers, and epidemiologists began to work on the evaluation, synthesis, and dissemination of trial-based evidence. On the basis of these syntheses, evidence-based protocols or "guidelines" for good practice could be formulated. These claims greatly appealed to health care policy makers, regulators, and managers looking for ways of keeping costs under control.

Not everyone shared this enthusiasm, however, and the concept of evidence-based medicine has been criticized by medical practitioners as well as by social scientists.[73] Some doctors stress that having to follow guidelines fails to respect the importance of clinical experience and judgment in treating an individual patient. People differ, and randomized trials express only the average benefits of an intervention applied to a group.[74]

Resistance on the part of the medical profession has to do with clinical autonomy, but it also has epistemological grounds. What counts as evidence, gets collected, analyzed, and published is a particular kind of reduction from the clinical encounter, and one that relies principally on statistical, rather than clinical, reasoning. Trials can yield results that directly conflict with clinical experience (as well as with the interests of clinicians). In the early 1990s, a Canadian National Breast Screening Study indicated that annual mammograms were of value to women in their fifties but not to women in their forties. The conclusion, that younger women should therefore not be offered annual mammograms, infuriated radiologists. In the medical and popular press, on television, and at medical conferences, radiologists attacked the quality of the study. There were even hints of fraud. Radiologists in turn were accused of trying to defend a lucrative practice. But more was at stake than status and income alone. Patricia Kaufert, a sociologist from the University of Manitoba, interviewed some of the radiologists. She realized that, for them, "being able to see and show where the tumor lies has an intense reality, besides which the epidemiologists' statistics on changing mortality rates become mere manipulations of a series of numbers."[75] In the United States, after a long and bitter struggle and against the advice of the National Institutes of Health, screening for younger women was restored by a vote in the Senate. In a purely medical arena, epidemiological evidence may have had the greater authority. But in a political arena, clinical—and personal—experience was the more powerful.

An additional complication is that disagreements regarding the value of a new medical intervention cannot always be resolved on the basis of evidence. Some new procedures, particularly ones bearing on the beginnings or ends of life such as artificial reproduction, prenatal genetic diagnosis, and organ transplantation, raise questions of values. Differences in values and beliefs lead physicians to conflicting views regarding, for example, the procurement and use of embryonic stem cells for research purposes, or preimplantation embryo selection. Evidence-gathering and synthesizing approaches like MTA and evidence-based medicine have avoided dealing with interventions like these. Policy making, determining what should be allowed and what not in cases like these, demands a broader consideration of the public good than medicine alone can provide. How and in what terms can this consideration take place? Where and by whom can controversy legitimately be resolved?[76]

Bioethical analysis was intended to help resolve these difficulties. Though there is a good deal of disagreement about the origins of bioethics, it is reasonable to say that, as initially conceived, it had "as its main task the determination, so far as that is possible, of what is right and wrong, good and bad, about the scientific developments and technological deployments of biomedicine."[77] Rights and wrongs would be debated in terms of the fundamental principles of "respect for persons," "beneficence" and "justice." Over the course of time, the philosophical complexities of these concepts were set aside. Bioethics was reduced to a set of simple principles that demanded no skills in philosophical reasoning and could be applied routinely. Thus, "respect for persons" became "autonomy," and "autonomy" became "informed consent."[78] The "stripped-down, monolithic version of principlism" that by the 1980s dominated bioethics in the United States and much of the rest of the world can be seen as a response by the emerging field to the cultural, legal, and institutional environment in which it was practiced.[79] The broader concern with "the likely effects of biomedical knowledge and its application on the human condition: the appropriate role of biomedicine in promoting human welfare and sustaining such important institutions as the family and community" was lost.[80]

Despite critiques, both evidence-based medicine and a "stripped down" bioethics continue to flourish in practice. Committees and working groups around the world continue to produce evidence-based clinical guidelines.

Bioethical committees are established, and bioethicists trained, in countries where such things were unknown until recently.[81] Medical historian George Weisz suggests that one reason for the continuing value of evidence-based medicine is the "protean" character of the concept, as easily invoked by politicians and health administrators looking to restrict clinical autonomy as by doctors looking to defend it.[82] This could apply to bioethical principlism, too.

These methods for evaluating the effectiveness of an innovative treatment tell us little about the considerations that led to its introduction. Since the weight given to statistical evidence as against clinical judgment varies from one decision making arena to another, while some technologies but not others will have been subject to ethical approval (with the very nature of that ethics evolving over time), it is hard to draw conclusions knowing only the outcome of the process. In other words, it is difficult for a patient to know what lies behind the advice of a specialist, advice such as Professor Smit gave us in 1989.

Turning back to the level of collective decision making, two further questions arise. Can ever-refined methods of analysis (based on trial data, medical literatures, or ethical principles) provide sufficiently credible and authoritative answers to the problem of choice? And, however thorough the analyses, are they in fact likely to determine what happens in practice? There is no easy answer to either question. Many STS studies have shown that the status of a piece of evidence or argument typically depends on the arena in which adjudication takes place.[83] When decisions are made within a medical arena and (ostensibly) on medical grounds, then rigorous and robust trial data, taking costs and benefits into account, should provide justification enough. But this depends on decision making being "contained" in an expert, medical forum, and not "leaking out" into politics or the courts. How realistic is it to expect decision making to remain contained, given fascination with new health technology, the commercial interests involved, and demands for wider participation in decision making? And what if this leakage (which has also been called democratization) does occur? Can the experts in analysis—the epidemiologists, statisticians, ethicists—"hold the line"? The question was put by William Gardner, a psychologist and bioethicist at the University of Pittsburgh, in 1995. Gardner asked whether it would be possible to prohibit parents from producing genetically enhanced children with desired characteristics, even if ethicists were agreed that the procedures should be prohibited.[84] He was doubtful: "Prohibition of genetic enhancement is likely to fail because it will be undermined by the dynamics of competition among parents and among nations." The more other parents do it, the more it becomes attractive to any single parent, both because they will wish to give their child this advantage and because they will by then be reassured regarding risks. As Gardner puts it, "Both nations and parents have strong incentives to defect from a ban on human genetic enhancement, because enhancements would help them in competition with other parents and nations."

When neither scientific evidence nor expert adjudication provides sufficient guarantee of widespread trust, assent, or conformity, then what else can be done? In the last few years, this question has been faced by a host of regulatory bodies in the field of health and medicine, for which gaining the trust of the public is crucial. Again, this is particularly apparent in the field of genetics, where pressures for change from science and industry are strong, but confront powerful though inchoate concerns and objections from civil society.[85] One way in which regulatory bodies in this field have adapted to the wide range of conflicting perspectives set out by patient groups, religious groups, industrial associations, and scientists is through developing mechanisms for public consultation. Seeking to establish their legitimacy through democratic debate, they have broadened their memberships to include members of the public and have developed consensus conferences and other mechanisms for consulting the community. Social relations and social processes are increasingly seen as crucial to sustaining trust in the governance of controversial biotechnologies.[86]

Dialogue and the search for consensus have become important in resolving controversy and deciding on an acceptable and legitimate course of action. There is an intriguing convergence between this development and how social scientists engage with the biomedical controversies that they study. Reflecting on their work, some found that it led them to question the sharply polarized positions they had seen expressed in the public sphere. As they explore how people live with technologies in practice, it is not controversy they find but ambiguity and doubt: "data from an increasing number of studies of social encounters with the new genetic knowledges and technologies reveals that few people, be they users or providers of genetic information, experience its challenges and demands in such 'either-or' terms."[87] It is this daily lived reality, not ideological conflicts in public, that should be the focus of social scientific inquiry.[88] Social scientists have found that their research practice confronts them with dilemmas of their own that are both moral and methodological. STS scholars have debated the question of whether they should "take sides" when studying a controversy involving science or technology. Some had found that despite trying to avoid taking sides in the controversy they were studying, they nevertheless found themselves embroiled in it.[89] They thought it had to do with the fact that the two parties to a controversy have unequal interests in a piece of social science research. Typically, they were studying situations where a group of "dissidents" challenged some established orthodoxy. The dissidents, typically weaker, try to get the sociologist on their side. Some sociologists of science argued that they have to resist attempts at "enrolling" them. Even though their work might later be used for political ends, in doing their research they had to try to stay neutral.[90] Sociologists and anthropologists, too, have debated the moral and the methodological pros and cons of "consciously intervening in an organization's social life."[91]

Researchers have dealt with these dilemmas in a variety of ways, depending in part on where and how they were working. Renée Fox and Judith Swazey were ultimately so shocked by the aggressive culture of transplantation medicine and by the way in which it drew resources from the real health care needs of deprived and fragile members of society that they decided to abandon their research. To continue, even as ethnographic observers of events, would have entailed too great a moral compromise.[92] Others have reflected on the relationships and the obligations built up in the course of their work. What precisely these were and how they were to be dealt with was a matter for moral and methodological reflection. Studying molecular biology laboratories and biotechnology companies, anthropologist Paul Rabinow recognized how much he had in common with the scientists he was studying.[93] He could not let them become pure objects of observation. But nor could he identify with them totally as fellow scientists, as soul mates. "The danger," he writes, "is in losing the balance, tipping too far toward the subjective side or the objective side. The ethical task is finding the mean."[94] How does one fulfill this "ethical task" of "finding the mean"? What makes it difficult has to do with the common human needs the ethnographer shares with his or her object of study, as well as the fact that those "objects of analysis are also analyzing subjects," with their own projects, values, and questions.[95] For other ethnographers in different settings, the dilemma can take a different form. Sue Estroff's work with people suffering from persistent mental illness also involved her in developing relationships with the people she was studying. But these were people who may have been "vulnerable, impaired, pained, and lonely."[96] Estroff was encouraged to play the role of "trusted messenger." The people she talked to hoped she would convey their views to those with authority over them in ways they themselves could not manage. Should she do what they wanted her to do? It was a fulfillment of the sense of obligation she felt, but it was also an "intervention" in what she was studying.

Though ethnographers disagree as to whether doing so is appropriate, some have made intervention an explicit element of their research strategy.[97] Deborah Heath did in the course of her of study of people suffering from Marfan syndrome (a genetic disease) when she brought researchers, clinicians, and patient advocates together in a public workshop. Moving between the worlds of laboratory, clinic, and advocacy groups, she obliged each to confront the understandings of the other in a way that was sometimes discomforting.[98] Lynn Sakai, a biochemist who figures prominently in Heath's account, was "unsettled" by thinking about patients' hopes and expectations. How could she reconcile her knowledge of those hopes and expectations with the value she placed on her scientific autonomy? Should we try to "unsettle" the scientists and clinicians we study? Some anthropologists are starting to say that yes, we should because what we are really unsettling is the status system of ways of knowing. The different ways of knowing an illness are embedded in different cultures in a hierarchical relationship to each other, so "intervening" through "mediation" is precisely what we have to do, argue Gary Downey and Joseph Dumit.[99] When Emily Martin and her collaborators wrote about immunity and the immune system, they treated the understandings of the research immunologists and injection drug users at high risk of (or infected with) HIV whom they talked to as of equal significance.[100] Explicit commitment to enhancing the visibility, or audibility, of the concerns of the stigmatized speaks loudly from this work. Martin and her colleagues discovered that they had broadened and deepened channels of communication across a cultural boundary marked by, and marking, inequality.[101] This "intervention through mediation" might be a way in which social scientists can help societies deal with the dilemmas that medical technology poses, and in a manner compatible with democratic ideals.

In carrying out this study, I too had to search for ways of finding the "balance" to which Rabinow refers. The tension that he and other ethnographers experience was made more acute by the fact that I was involved both as parent and as researcher. The duality of my role presented me with additional dilemmas: for example, deciding what of myself to reveal to people I was about to interview. I suspected that clinicians would respond differently to questions posed by a social scientist versus the father of a deaf child. How much to reveal, and when? The dilemma was indissolubly moral and methodological. As a parent, I had insight into the needs of deaf children and into the adequacy of the provisions made for them independent of my research. But as a parent, I also had a commitment to my children's well-being. In how far and in what ways could I allow these personal insights and interests to influence my interpretations of what I discovered? Ethnographers try to deal with obligations built up in the course of their research. That was true for me, too. The difference was that I began this research very aware of prior obligations to my family.

Since "evidence" figures so largely in current health care discourse, in carrying out this study I paid particular attention to the evidence for the effectiveness of the implant: what counted as "evidence" and how it was used at each stage of development. How far did the experience of deaf people themselves, their satisfaction or dissatisfaction with the implant, count as evidence, and at what stage? The technology evolved from the "tinkering," in the 1950s, through the beginnings of commercial manufacture, to the gradual establishment of medical consensus regarding the value of the implant for deaf adults in the early 1980s. By the end of the 1980s, ear surgeons were starting to agree that it could benefit deaf children, too (though in 1989 Professor Smit had not yet been convinced). I found that ethical considerations also played a complex role in the history of the implant. Ethical arguments were invoked by critics questioning the value of pediatric implantation in particular, and ethical committees sometimes played a role in attempts at adjudication. However, the interplay between different views of the utility of the implant, and the roles both of evidence and of ethical deliberation, differed from one country to another. I tried to understand how and why that was. Starting in the early 1990s, cochlear implantation was offered to deaf children with increasing frequency. Their parents had to decide what to do just as we had had to decide. I tried to understand on what basis they did so, and how parents sought to weigh the evidence and arguments in a way consonant with today's discourse of empowerment. The discussion in this book thus addresses the three broad themes mapped out in this introduction. Where did the technology come from and how did it become a source of hope? Were, or are, there alternatives? How can we know what is best: as parents, as patients, and collectively as societies? In the book's final chapter, I discuss the implications of this history of the cochlear implant for issues of evidence, of rights, and of democratic decision making that have today become problematic features of new medical technologies more generally.

Chapter 2

The Making of the Cochlear Implant

On February 25, 1957, Professor Charles Eyries of the medical faculty in Paris became the first surgeon in the world to try to give a deaf patient some hearing by means of an electrode implanted into his ear. That hearing was associated with the passage of electrical currents had been established long before. As early as 1800, the famous Italian physicist Alessandro Volta (after whom the volt is named) placed electrodes in each of his ears and connected them to a battery. The electrical current that passed through his head produced what he described as a "disagreeable sensation," and a noise "like the boiling of thick soup." Other scientists tried to repeat his experiment. Some succeeded and others failed. But the seeds of research on the electrical nature of hearing had been planted.[1]

The currents involved are tiny, and it was only when development of the electric vacuum tube made it possible to amplify them that they could really be studied. In the 1930s, much of this research made use of animals such as cats and guinea pigs. For example, one experiment involved using a live cat as a microphone. If electrical contact was made with the cat's auditory nerve and the signals developed in it amplified and passed through telephone receivers, a listener could recognize the sounds delivered to the animal's ear. If words were spoken into the cat's ear, the listener could identify them.[2] The relation between an electrical signal and the perception of sound to which it gave rise became a major focus of research. For this, human volunteers were needed, since experimental animals obviously could not describe what they heard. A group of Harvard scientists applied various AC and DC currents to the ears of volunteers, with the live electrode in the ear and the ground electrode attached to the arm. The volunteers had to describe what they heard when a current passed. Though some reported hearing nothing at all, most heard something. What exactly they heard depended on their ears and on specific characteristics of the electrical circuit. As with the earlier cat experiment, it was also possible to bypass the subject-listener by connecting the electrodes to the output circuit of a radio. Then the researchers themselves could listen in. One of the scientists involved explained what they heard: "Music can be heard and popular tunes identified, but the quality is definitely poor—'tin pan' music. Speech can easily be recognized as speech, but only occasional words can be understood. Clearly, electrical stimulation does not promise much as an alternative means of hearing so long as so much distortion is present."[3]

By the end of the 1930s, it was known that the inner ear, or cochlea, functions as a transducer: that is, it converts mechanical sound waves into electrical signals carried by the auditory nerve to the brain. Understanding how exactly this works, how electrical signals yield the complex we understand as speech or music or just noise, lay far in the future. And although this was basic physiological research with no immediate medical application, the possibility of applying it, of helping people who could not hear, was in the background. Reminiscing decades later on the 1930s research at Harvard, one participant recalled, "Dr Davis and I gave the first demonstration at the international meeting of the Physiologists. The newspapers obtained a report of the presentation. The first thing I knew I received letters from individuals all over the world asking when could they come and have their hearing restored and that is the great danger of this work appearing in the newspapers and the ensuing publicity . . . There will be people demanding that these procedures be done on them."[4]

Early in 1957, a man suffering from a progressive infection of the middle ear known as cholesteatoma approached Eyries. Despite operations on both ears, this man had become totally deaf: a not uncommon result of this disease. At that time, there was little that ENT surgeons could do for people who could not hear. Great strides in the precise measurement of hearing loss had taken place, and electrical hearing aids had been developed in the 1940s. The amplification these provided was useful to people with a little hearing but did not help those who had none at all. Especially when the inner ear, which turns sound waves into electrical signals, was affected, doctors could do nothing. Research on the electrophysiology of hearing was advancing but had as yet yielded no practical benefits. Eyries's patient, in desperation, "expressed the desire that the impossible be tried in order to put an end, however imperfect, to his total deafness."[5]

Eyries approached his colleague André Djourno, who was studying electrical stimulation of the auditory nerve in animals. After some deliberation, they decided to try to implant the patient with an electrode similar to that used in the animal research. This would stimulate his (functioning) auditory nerve. The electrode was constructed by one of Djourno's assistants in the physics laboratory of the Paris faculty of medicine, and on February 25 it was implanted into the patient's right ear.

The next few weeks were strange but enjoyable for the implantee, and even if he was unable to follow speech, he did enjoy hearing some of the sounds of daily life: coming and going, the bang of a door, the noise of conversation. The sounds he heard were strange, metallic, unlike what he remembered. In July, one of France's most renowned speech therapists, Suzanne Borel-Maisonny, started to work with him to help him learn to use his new prosthesis. But as his expectations failed to be born out, the patient's initial enthusiasm faded. He soon discovered that however hard he worked at it, he was not able to distinguish different kinds of sounds. Speech, opening a door, dragging a chair: they all sounded the same. In March 1958, the patient decided that he had had enough: it was not worth the investment of so much time and emotion. But if he was disillusioned, Eyries and Djourno were not. They tried twice more, the last time in November 1958. And though success—in the sense of providing a patient with hearing—eluded them, they were sure that it would one day be possible.

A couple of years later, a young Los Angeles ear surgeon, William House, recently qualified to practice, heard about what Eyries and Djourno had done via a newspaper clipping brought to him by a patient. He sought out the original article and had it translated. Intrigued, House constructed a simple electrode that he tried out on volunteer patients undergoing ear surgery.[6] He placed an electrode—an electrical wire—at the edge of the inner ear, passed a signal through, and asked his patients to describe what they heard. The next step was to find some volunteer patients who had lost their hearing and place electrodes in their ears. Did they experience anything like hearing? Satisfied by the results of these trials, which suggested that subjects had indeed "heard" something, in 1961 William House decided that he would now *implant* a more complex electrode into a volunteer: a man deaf from advanced otosclerosis. House's new implant was designed to stimulate the cochlea at five different positions along its length. This phase of House's work did not last long because of technical problems. The wires were insulated with silicone rubber that at that time contained some toxic substances. The implanted patient began to develop symptoms that led Dr. House to remove the electrode after about three weeks.

The problems were not only technical. The engineer who had constructed the electrode that House used became highly enthusiastic at the commercial possibilities of the new device. He informed the press, and House and his colleagues were deluged by calls from people who had heard about their work, just as an earlier researcher had anticipated. House was not ready for this. Convinced of the potential of the technique but alarmed, he decided to dissolve the partnership and stop work on cochlear implantation.

Some scientists were horrified by the newspaper reports of House's work. Blair Simmons, professor of otology at Stanford University Medical School, was one. He was enraged by what he saw as "irresponsible claims" that the deaf could be made to hear.[7] But with his long-standing scientific interest in physiological processes of sound reception, which had involved him in years of animal work, Simmons was also intrigued. Like Eyries and Djourno earlier, in 1962 Simmons had the opportunity to try something similar on a volunteer patient (who felt that he had nothing to lose). Simmons was an experimentalist who wanted to help resolve a theoretical debate in hearing research. This concerned the relative importance of two mechanisms by which a listener might perceive differences in the pitch of a sound. Some thought that this was due to differences in the rate of electrical stimulation ("rate encoding"), others that it had to do with where along the length of the cochlea stimulation occurred ("place encoding"). Simmons thought that the design of a cochlear implant would depend crucially on which theory was correct. He wanted to know if his deaf subject would perceive changes in pitch when the rate of stimulation applied to the electrode was changed. If so, this would be evidence for the "rate encoding" theory.

Simmons was encouraged by his initial experiment and by the mid-1960s had decided that he would try to develop a permanently implantable device.[8] To proceed, he needed to collaborate with an expert in auditory psychophysics or speech coding. This proved to be a problem. Even as an established Stanford professor, Blair Simmons failed to find anyone willing to work with him. As he later described the problem, the extravagant claims being made "both at clinical meetings and in the newspapers" had made many basic scientists hostile to the very idea. Talk of deaf people using the telephone, enjoying the chirping of birds put basic scientists off. They did not want to become associated with an enterprise that might damage their reputations in the eyes of fellow scientists. Nevertheless, in May 1964 Simmons and a Stanford colleague implanted a six-electrode device in a sixty-year-old volunteer who was totally deaf in his right ear and was losing his hearing in the left. The subject was also losing his sight due to retinitis pigmentosa. "We were amazingly lucky," Blair Simmons commented later. "All electrodes functioned and remained so until he was explanted eighteen months later." The subject learned to tell whether what he was hearing was speech or not, but he could not identify individual words or phrases. The tests done with him did not take place at Stanford but at Bell Labs in New Jersey where one of Simmons's graduate students had contacts. Scientists at Bell were willing to collaborate whereas those in California with similar expertise were not. Why was this? Simmons had an explanation: "They were outsiders. I don't think they'd read the publicity."

The result was that this elderly man, nearly blind, had to be taken to New Jersey for independent confirmation of what the Stanford studies had suggested. He had never been on a plane before, and because neither Bell nor Stanford was willing to assume responsibility for his well-being during the journey, Simmons and his wife traveled with him. All went well, and the scientists at Bell Labs obtained the results that Simmons had hoped for. But the scientific and medical communities were unconvinced. The American Otological Society refused to allow him to present the work at their 1965 meeting, while an application for funding was turned down by the National Institutes of Health. Simmons turned back to experimental research on cats. Human auditory stimulation, it seemed, was not an area in which you could do experimental research.

Changing Climate of Opinion

That attempts to develop a cochlear implant were resumed in the late 1960s was a consequence of technological advances and policy changes wholly extraneous to the field of hearing research. It was the heart pacemaker that gave the notion of an implantable device legitimacy and appeal. Initially developed for use in the operating theatre, the pacemaker only later became a permanently implantable device thanks to subsequent advances in electronics and in materials science.[9] In 1966, the new health insurance plan for the elderly in the United States, Medicare, agreed to reimburse pacemakers for people over sixty-five, with the result that the number of pacemakers grew rapidly. Three years earlier, with enthusiastic support from Congress, the National Institutes of Health had established an artificial heart program. In 1967, with the move of Willem Kolff, Dutch-born inventor of the artificial kidney, from the Cleveland Clinic to the University of Utah, the race to develop an artificial heart began in earnest. This work, in which Kolff, Michael DeBakey and Denton Cooley in Texas, and Adrian Kantrowicz in New York were the leading contestants, attracted enormous publicity.[10] In 1968, a research group in London suggested that it might be possible to develop an artificial retina for blind people. Kolff, perhaps the most inventive and determined of all artificial organ pioneers, was determined to make his institute in Salt Lake City the world's leading center of artificial organ development.[11] The physiological and engineering problems involved in artificial eyes and ears were seen as similar, and Kolff's group started to investigate both. Hearing research profited from the much greater popular enthusiasm that the idea of an artificial eye seemed to provoke.

It was these changes in the research and policy contexts that inspired Simmons and House to turn back to cochlear implantation. Their immediate objectives and styles of work were still very different. Simmons wanted to develop the implant he thought deaf people would need if they were ever to understand speech. It would be a long-term undertaking: designing such an implant would require a great deal of research, both neurophysiological (using animals) and electronic. With the aid of a grant from the NIH, Simmons started on this research. House approached the problem differently, more as a clinician. Convinced that the principle had been proven, he wanted to find a way of putting the technology to use as quickly as possible. There were patients desperate to be helped, and he felt it his responsibility to help them.

The coexistence of such contrasting approaches is not unique to this field. When Renée Fox and Judith Swazey studied the early years of transplantation surgery, they had the sense that transplantation surgeons divided into two groups. Despite a common commitment to pushing medicine beyond its existing frontiers, "only some of these men wield their microscopes as often as their scalpels, and work with their laboratory dogs as much as with their patients."[12] These "experimentalist surgeons," as Fox and Swazey call them, are more inclined to caution than their "clinician" colleagues, more inclined to work initially with a limited number of "well-studied patients," drawing as much as possible out of their data. In the case of cochlear implantation, relations between the experimentally oriented and clinically oriented innovators were far from cordial. In other respects, the process through which the cochlear implant took shape had a great deal in common with the development of other medical technologies in the same period. There is a clear pattern of development. A technically inclined physician or surgeon, recognition of a medical need, an initial "tinkering" trial-and-error phase, collaboration with a local engineering company, skepticism from colleagues: the pattern is much the same in the histories of diagnostic ultrasound, of the intraocular lens, of the artificial hip, and others.

For House, anxious to have a device available for clinical use as fast as possible, the need for an industrial partner was far more urgent than it was for Simmons with his long-term research agenda. Even though there was no funding, Jack Urban, the president of a small engineering firm with interests in medical electronics, was willing to construct an implant for him. In 1969 and 1970, three patients were implanted with a five-channel device constructed by Urban's company. Signals and power were transmitted via a transcutaneous button (a kind of small plug set into the skin behind the ear). In one case, there were medical problems—the implant was rejected and had to be taken out—while a second patient moved out of the area. Substantial work was done with the third patient, Chuck Graser, a deafened high school teacher. The implant and the "speech coding strategy" (that is, the way in which speech or selected speech frequencies were presented electronically) were gradually improved. In a paper presented at the American Otological Society in 1973, House and Urban quoted at length from a letter Graser wrote to them:

> You would probably describe my current progress as changing from profoundly deaf to just hard-of-hearing, but difficulty hearing and comprehending is in a completely different league from silence. For instance, tonight I can finally hear the bell that indicates I am at the right hand margin, as I type this letter.
>
> A year ago I heard nothing unless I was at Urban Engineering with test instruments. Tonight I went bowling and heard all the noise of people's voices, pins falling, clapping and so forth.
>
> I can call home from work and ask my wife what she is planning for dinner, or what she bought on her shopping trip. I also get tremendous help and enjoyment hearing a blue jay squawking, the cat meowing, a twig breaking underfoot, bacon frying, the door bell or phone ringing and so on.[13]

William House became convinced that he could help people like Graser without spending years developing the complex system that Simmons thought necessary. Why work endlessly to develop a complex multichannel device when a much less expensive single-channel one could do so much? House and his colleagues changed their strategy and set about developing a program of rehabilitation so that patients could learn to exploit the limited possibilities of the single-channel device. By October 1977, House had implanted devices in twenty-two patients, compared to Simmons's two or three experimental subjects.

By the mid-1970s, other research groups had gotten involved. One was at Kolff's Institute of Biomedical Engineering at the University of Utah. Another was at the University of California at San Francisco (UCSF) where Dr. Robin Michelson (a physicist turned otologist) implanted four patients. Not only were relations between the groups sometimes far from cordial, they disagreed profoundly about what had been achieved and what had yet to be done. A workshop in San Francisco in 1974 showed disagreement about the experimental or "experimental/therapeutic" status of the technique (with most inclining to the former view), about the proven (or not) value of the single-channel implant, and about the additional benefit to be expected from a future multichannel device. Despite the many differences of opinion, there was one point that attracted general assent: prospects for the technology were very favorable. Of the 300,000 profoundly deaf individuals in the United States, "as many as two thirds . . . might derive some potential benefit from an implant device."[14]

A couple of years after Michelson had started implant work, Michael Merzenich, a physiologist, moved to UCSF. Appointed jointly to the departments of ENT and physiology, Merzenich had no desire to get involved with cochlear implantation. The head of the physiology department advised him to keep away from it. Moreover, he was convinced that Michelson's understanding of what the implant was to do was erroneous.[15] Michelson's requests for Merzenich to see one of the implant patients were resisted, until finally he agreed on the condition that he could himself test the patient's hearing. As he later explained, "I was flabbergasted. I was amazed what this person could hear and distinguish with the single channel. I knew enough about Vocoders, about speech representation, about speech processors to know that, if you could obtain that much information from a base channel then there was a possibility that you could represent speech . . . I more or less instantly got the bug. I thought this was something that should be studied."[16] Merzenich decided that he would collaborate with Michelson. He set up a program of physiological research that should lead to the design of a multichannel implant system.

Apparently skeptics could be convinced, but there was still a good deal of opposition. Many clinicians still feared the exaggerated press reports and unrealistic expectations to which cochlear implantation work could give rise. William House recalls that colleagues were not keen to have him present his work at national meetings at that time.[17] Many scientists working in the physiology and neurology of hearing felt that too little was known for this kind of human experimentation to be justified. While speech recognition via electrical stimulation ought in principle to be possible, though never with a single channel implant, major surgical and physiological problems remained to be solved. Nelson Kiang, a leading figure in the physiology of hearing, emphasized that too little was known of how the central nervous system processed sound for prosthesis design to be based on anything more than guesswork. Additional animal studies were essential before clinical work was taken further. Kiang was profoundly opposed to what a number of clinicians were doing. When House and Urban presented their work at the 1973 American Otological Society meeting, Kiang delivered a scathing attack on it: "Dr. House's results are no different from those of previous workers except that the criteria applied to the definition of success have been lowered. Enthusiastic testimonials from patients cannot take the place of objective measures of performance capabilities."[18] House responded simply, "I am trying to do everything I can to improve the patient's situation."

Responding to Kiang at that 1973 meeting, a representative of the Utah group also got up to speak. "I think a recent remark made in my presence by 'Pim' Kolff, inventor of the artificial kidney, is very important and bears repeating. When asked about the fact that, after thirty years, the artificial kidney was still not fully understood, he replied, 'If I really worried how it worked, I would still be studying membrane transport in cellophane, instead of building the first artificial kidney.' I feel the same way about the auditory prosthesis. If it works, I will take it. Auditory physiologists like you, Dr. Kiang, can then try to explain why."[19]

Having been a California phenomenon in the 1960s, in the 1970s development of a working auditory prosthesis was becoming an international activity. William House's work in particular was attracting the attention of clinicians in Europe and Australia. Many of them listened with fascination to his lecture at an international conference in Venice in 1973.

Perhaps an old dream lay at the root of their fascination. Would it finally be possible to give hearing to the deaf? This was a challenge that had so long resisted the desires and efforts of the medical profession. Some of the pioneer researchers refer to dreams of that kind. One writes of having "dreamt constantly of an electrical system, a James Bond-style gadget, which would be able to alleviate the formidable handicap of total deafness."[20] Another was said to have been "haunted" by an "old dream of restoring hearing to deaf people."[21] But how could these dreams be turned into practice? What lessons could be drawn from what House, Simmons, and the others had achieved? Differences of style, of temperament, of commitment to treatment or to science, differences that had been there since the start, remained. There were some who took a direct route, like House had done, trying to help patients clamoring at their doors. Claude-Henri Chouard in Paris was one of these. Others, like Simmons, were convinced that more research had to be done before a working prosthesis was possible. They would start by doing this more basic research. Ellis Douek in London was one of these, Graeme Clark in Melbourne another.

Claude-Henri Chouard, an ear surgeon at the Saint Antoine Hospital in Paris, had been a student of Charles Eyries. He knew of Simmons's and House's work. He had visited Los Angeles and seen House operate. Chouard was rapidly convinced that people would be able to hear with such a device. "What interested me was making people hear and not making animals hear," he explains. Since he was sure he could help people, he saw no need to work on animals. In May 1973, Chouard went to the Venice conference with a neurophysiologist, Patrick MacLeod. Having listened to House and to "two Americans from San Francisco who'd been working mainly on animals," Chouard and MacLeod began to think about how an implant that could give deaf people language would work. Through the summer of 1973 they worked on their design. The theory of place coding suggested that if they could stimulate the cochlea at a sufficient number of points along its length, some frequency discrimination should be provided. This, they reasoned, was essential for the understanding of speech. By the end of 1973, three people had been implanted with their new multichannel device. Chouard became more and more enthusiastic, convinced that his dream was on the verge of becoming reality. To go further, however, he needed more money for his work.

He could not rely on the support of his professional colleagues; far from it. This meant he had to think carefully about where he could find the funding he needed. Risking the wrath of medical colleagues, he deliberately brought his work to the attention of press and television. The Ninth International Congress of Audiology, which took place near Paris in May 1974, provided an opportunity for doing just that. The implantees (patients) had an important role to play: "The press, the radio, the television, fond of the spectacular, make the discovery known, explaining what needs still to be done." "My morale is of steel," he recalled later. "In front of the television, my patients have become 'performers' . . . We make them work relentlessly, to teach them to recognize sounds, words, phrases, so that they would be able to appear before the DGRST [Délégation Générale à la Recherche Scientifique et Technique] commission which would have to grant us our subsidy."[22]

As he had intended, in April and May 1974 Chouard's work received considerable publicity in the newspapers and on television. The message the media carried was a relatively simple one, stressing the themes of "deafness vanquished" and "French triumph." Headlines such as "Hope for 2 Million Deaf and 17,000 Deaf Mutes"[23] or "Victory over Total Deafness"[24] did little to endear Chouard to his colleagues.[25] But the strategy paid off: Chouard received that vital subsidy.

The next problem was to find an industrial collaborator to produce an implant based on his design. The firm with which he collaborated, Bertin, was able to make some important improvements. They replaced the teflon plug into the head that he had been using with an electromagnetic coupling, a kind of wireless transmission. This required an antenna attached behind the ear to a spectacles frame and a receiver implanted in the mastoid bone. A miniaturized device implanted into the skull decoded the messages and passed information to each of the electrodes. The total prosthesis, named the Chorimac, consisted of an externally worn box, an antenna, and the implanted receiver and electrodes. The external box, weighing a massive 2.3 kilograms, contained a power source that could be recharged nightly, a microphone, a compressor (to keep the volume of sound within the right limits), and filters (each of which allowed sound in a particular range of frequencies). In late 1976, the first three prototypes were ready.

By now convinced that the value of the technology had been proven, Chouard performed as many implant operations as his resources allowed: roughly one a month throughout the mid-1970s. He was not deterred by the fact that French neurophysiologists, like their American colleagues, were highly skeptical. In 1977 he began to implant children, something which neither House nor anyone else was doing. A group of visiting British experts was surprised at this: "The fact that there are apparently no problems about implanting juveniles and children in France means that, in time, information should become available which, at present, cannot be obtained elsewhere."[26]

The rate of implantation, the publicity that his work was attracting, the implantation of children: all these things provoked hostility in French medical circles, though few were prepared to voice their objections in public.[27] One who was, was Michel Portmann, director of the ear, nose, and throat department at the University of Bordeaux. As far as he was concerned, trying the device out on children was simply unacceptable at that point in time. Nor did he think that Chouard had made the right design decisions. Together with colleagues in Bordeaux, Portmann was carrying out his own experiments. These would later lead to their own prototype single-channel extracochlear device, the PRELCO—one of many designs that were to fall by the wayside. But Portmann was a major figure in French otology and his criticism mattered.

Years later, Chouard explained the opposition he had faced as a normal aspect of the acceptance of novelty in medicine.

> The doctor is conservative in the sense that his duty is to conserve life. And every new therapy seems to him an aggression that might be dangerous. Think about the inventors of X-ray. They burned their fingers because they didn't know. Resistance is normal. That opposition embarrassed me a great deal in the beginning, but I have always respected my opponents, and in particular I continue to understand the reticence of some people who work with deaf children and who still continue to believe that for the young child the cochlear implant is a bad thing. But you know, these objections . . . evolve in three steps. The first consists in saying, "What he is saying is false, it isn't true. He's lying." The second step consists in saying, "Yes, it's true. What he's saying is true, but it is without significance." And the third step, and that's where we are now, and which I am relishing with a great deal of pleasure, is to say, "Ha, it works . . . just like I always said." There are those who opposed the project and who now say that it is fantastic and must go forward. In France there are many schools that talked like that

and which are now avid defenders of cochlear implantation. I suffered from this way of doing things, but it is quite normal. I would even say it is physiological. There you are.[28]

In Britain, the start of cochlear implant research and development was accompanied neither by any James Bond-style dreams nor by opposition such as Chouard faced. It was a much more low-key affair. It all started in the early 1970s, soon after Ellis Douek's appointment to a senior ear, nose, and throat post at London's Guy's Hospital. The Department of Health, prompted by a deafened Member of Parliament active on behalf of the disabled (Jack Ashley, now Lord Ashley), suggested to Douek that his specialty was doing far too little on sensorineural deafness, and why didn't he do something in that area? They suggested he apply to the Medical Research Council (MRC) for funding. Douek decided that, before visiting the United States and seeing what House and the others were doing, he would try it himself. This he did: sticking an electrode to the outside of a patient's cochlea. The results—which seemed to show that you could get the same effect House got with an *i*mplanted electrode but without the surgery—came as a surprise. It was enough, it seemed, to attach an electrode to the entrance to the inner ear (the so-called round window). Douek went to the MRC, but the council's scientists were not impressed by his rudimentary experiment. Like their colleagues in the United States, they were skeptical. Real language was far too complex for such an approach to work.

The MRC put Douek in touch with Adrian Fourcin, an expert in phonetics at London University. "We went to his department, which was like a magic cavern for me, with equipment that I'd never seen, and so on. He had an apparatus, which he had invented, called a Laryngograph. He said, 'Look, if we put these electrodes on someone's neck it will record the changes in the pitch of the voice. Not speech. Speech is the mouth' . . . I'd thought of speech and voice as one thing . . . He said to me 'Look at this.' And there was an analysis of all the voice recordings with the Laryngograph. He said 'What does that remind you of? Isn't it exactly the same pictures that you were able to produce by electrical stimulation?'"[29] Fourcin's idea was that you could use an implant to help the deaf lip-reader. There are many sounds in speech that are very difficult to distinguish solely by their appearance on the lips. This is where the implant would come in. It was quite a different objective from that of trying to provide access to speech directly, without lip-reading, as others were trying to do.

In late 1974, the MRC set up a working group to review what had been achieved up until that point and to recommend what further research was needed. Both Douek and Fourcin were members. The working group concluded that while artificial stimulation was potentially useful "to the small number of patients who become totally deaf through a cochlear degenerative disorder," its practical achievements to date were limited.[30] A simple approach was recommended, involving placing a single electrode on the round window of the cochlea. Avoiding the risks of implantation and the complex electronics of other approaches, further work along these lines should provide essential information on pathology, physiology, and psychoacoustics. In 1976, the working group's recommendations were accepted by the MRC. Douek and Fourcin, together with Cambridge psychologist Brian Moore, worked out a research proposal. It was approved for funding, and in January 1977 their project started. It involved the Hearing Research Group at Guy's Hospital (including the surgeon Douek, a medical physicist, and an audiologist); the Department of Phonetics at University College (University of London); and the Department of Experimental Psychology at Cambridge University.

Douek and his colleagues saw their project as different from what others were doing. First, deriving from Douek's initial experiment, was the extracochlear approach. This involved less surgical risk to the patient and would not destroy any hearing in the implanted ear. After all, the device could fail to work, and then the patient wouldn't have lost anything. Second was Fourcin's important contribution that, at least at first, the implant should be used to supplement the information available from lip-reading. Anyone who became deafened in adult life would remain dependent on lip-reading. The first goal was to add to the information that could be "read off" from the lips. The attempt to provide "hearing," they saw as unrealistic, at least at first. This choice had important technical implications. It meant that they would try to transmit not the whole speech signal but a selected sample of the sound frequencies that make up speech. A "speech processing strategy" would be needed that tried to compensate more precisely for limitations of lip-reading.

Like Blair Simmons, the London-Cambridge group had no intention of implanting large numbers of people. They saw themselves as a research team. Douek, Fourcin, and Moore were not looking for publicity like Chouard was. Since they were not offering a clinical service, there was no need to attract large numbers of patients or to demonstrate the effectiveness of the device in routine use. They had all the funds they needed from the MRC. Like Simmons, their objective was to get as much information as possible out of a small number of experimental subjects. These could be found without appeal to the mass media.

Another pioneer was the Australian Graeme Clark. Clark's motivation to become an ear surgeon, to help the deaf, was clearly influenced by his experience growing up as the son of a deaf father.[31] Having completed his training and entered practice, in late 1966 Clark came across an article by Blair Simmons:

"a profoundly deaf person had heard some strange sounds when electrical currents were passed through wires placed on his hearing nerve. Although the person couldn't understand speech, that initial report was enough to fire me up. I felt this was my mission in life."[32] Clark decided to abandon his surgical practice in order to devote himself to research. Though he soon knew of House's work, Clark felt, like Douek, that too little was known about how sound was processed in the brain. How could you put in an implant when so little was known about how it would have to work? Between 1967 and 1970, Clark busied himself with a PhD thesis. Using animals, he explored how sound is encoded and transduced into nerve impulses.

On the basis of this work, Clark was convinced that stimulating the hearing nerve with a single electrode—making use of "rate coding" alone—would not be enough to provide access to speech. For this purpose, multiple electrodes inserted at different positions in the inner ear—allowing use to be made of "place coding" as well—would be essential. Like Douek and Fourcin, Clark was convinced that selected elements of speech, rather than the whole speech signal, should be fed to the brain. So a processing strategy was needed to select out the right parts of the signal. What should it be? These were the kinds of things he would need to know in order to design an implant. Clark was then working in a department of physiology. He was sure that this was not the place to pursue his dream. Physiologists in Australia, as elsewhere, "thought it was outrageous to suggest that electrical stimulation of the inner ear could adequately reproduce frequency information to help profoundly deaf people understand speech."[33] In 1970, appointed to a chair in otolaryngology at the University of Melbourne, Clark set about establishing the research program he would need in order to design a multichannel cochlear implant. According to the "place" theory, he reasoned, nerve cells sensitive to the range of frequencies necessary to understand speech would have to be stimulated along the length of the cochlea. How could all the wires be inserted in this tiny spiral-shaped organ? Invoking a familiar "Eureka" image, Clark writes of wrestling with this problem day and night until one day in 1976, playing with a shell on the beach, an idea flashed into his mind. A blade of grass, pushed into the opening of the shell, bent easily around the spiral. All depended on the strength and flexibility of the electrode.

For Clark too, funding was initially a problem, and opposition from scientist-colleagues did not help. Like Chouard, Clark felt that the answer would be found in appeals to the public. For him too, publicity would therefore be vital. Fundraising became a major preoccupation. A donation of two thousand Australian dollars from the Melbourne Apex club led to a small item on a news broadcast. One who saw it was business tycoon and television channel owner

Sir Reginald Ansett. Interested in the problems of the deaf because his daughter had a deaf friend, Ansett launched a telethon to raise money for research on nerve deafness in 1973. For four years, all the money raised went to support Clark's research: more researchers, better equipment, and more publicity. Additional funding, in the form of grants from the National Health and Medical Research Council and from Lions Clubs International followed. Despite professional opposition similar to that faced by the other pioneers, Clark was securing the resources needed to push his research further.

By 1977, they were ready to try their prototype device on a volunteer. But again, it took time and publicity to find a suitable volunteer. On August 1, 1978, a prototype was implanted into the first volunteer, Rod Saunders, who had lost his hearing in a traffic accident a couple of years earlier. The operation was hailed as a great success. But what exactly could Saunders perceive with his implant? Could he recognize tunes, for example? To the researchers' delight, it turned out that he could. In September, Clark felt confident enough to announce their success to the press. There was still a long way to go, however. Their speech processor was not yet portable. Rod Saunders could only hear with his implant when connected to an immovable wall of computers. It was to take a further two years before the portable processor was ready.

By this time, cochlear implants were being developed by a dozen or so groups scattered around the world. Each of them had had to assemble a variety of scientific, medical, technological, and manufacturing skills, as well as financial resources and—crucially—volunteers to receive the implants. The mass media were sometimes mobilized in pursuit of these resources. This process, by which innovation gradually becomes vested in a whole range of agencies, knowledge, skills, and resources, was characteristic of the development of new medical devices through much of the second half of the twentieth century.[34] Success depended crucially on being able to assemble the necessary skills and resources, but what this entailed differed from country to country and from center to center.

The Networks Are Extended

By the end of the 1970s, cochlear implantation was gaining credibility in the ENT profession, and slowly but surely this credibility was being converted into support. In the United States, clinicians had been reassured by an independent assessment of House and Michelson implantees that showed modest but definite benefit.[35] As medical specialists let themselves be persuaded that the technology worked, wider social interests came into play. Large industrial corporations, eyeing potential commercial opportunity, began to take an interest.

So did public bodies concerned with the planning and finance of health services. In 1977, the British Department of Health (DHSS) asked three experts (John Ballantyne and Andrew Morrison, both ear surgeons, and Edward Evans, a neurophysiologist) to review current efforts in the field and make recommendations as to what British commitment should be. In October 1977, Ballantyne, Evans, and Morrison toured the implant centers in the United States. While critical of certain aspects of what they saw, the British experts were persuaded of the promise of the technique.[36] Viewing full-scale clinical provision as premature, they recommended a cautious approach in Britain, starting with a careful evaluation of the single-channel implant. This was valuable support for Douek and Fourcin.

Industry was also becoming interested. Links with industrial corporations become important as a new medical technology is adapted to medical practice in the real world, and become vital as it is prepared for the commercial market. In the case of the cochlear implant, these links were emerging in the period from 1978 to 1982. But they did not emerge easily. There were many false starts.[37]

While Douek remained uninterested in links with industry, some of the other pioneers were very interested indeed. Two firms played an important role at this time. 3M, the Minnesota-based multinational with a large life sciences sector, was the first company to market a cochlear implant in the United States.[38] In 1977, the firm's Australian subsidiary was approached by Graeme Clark. 3M management was interested, foreseeing a substantial market and concluding that "the first company to commercialize a device could possibly dominate the market, as had been the case in the pacemaker industry."[39] Before entering into an agreement with the University of Melbourne, 3M looked into what other work was going on. Contacts were established with William House in Los Angeles, with the two groups in San Francisco (at Stanford and UCSF), and with the Irwin and Ingeborg Hochmair husband-and-wife team in Austria. Contacts with the Australians came to nothing. The Australians were working on a complex 22-channel device that 3M people thought "too sophisticated."[40] They were not convinced that this was the way to go at that time.

Although they were looking for a simpler device, 3M pursued a complex development strategy. In 1981, the firm embarked on contractual agreements with William House and the Hochmairs. Their idea was first to obtain Food and Drug Administration (FDA) approval so that the House device could be put on the market. In 1976, the FDA's mandate had been extended from drugs to medical devices. For implanted devices like this one, the requirements for approval were stringent. After 1976, it was forbidden to offer a device like this for sale in the United States without first satisfying the FDA that it was safe and that it worked. At the same time as it sought FDA approval, 3M would work on "second generation" devices and begin to explore possible "third generation" implants. The first task was thus to prepare the House implant for submission to the FDA.

In the course of 1982, 3M made some important modifications to what House and Urban had designed. They reduced the length of the electrode that went into the inner ear from 15 millimeters to 6 millimeters. That would reduce the risk of damage to the ear during surgery. They also collaborated with the University of California at San Francisco (UCSF), where Michelson and Merzenich were developing a multichannel device. It was unclear which design would be better, and there was obviously something to be said (from the company's point of view) for not putting all of its eggs in one basket. However, the partnership with UCSF was short-lived. Two or three people were implanted with the prototype that Michelson and Merzenich had developed before the partnership broke up. 3M also started collaborating with the Hochmair team in Vienna, who were working on an extracochlear device. By late 1982, more than two hundred people with a sensorineural hearing loss had been implanted with House's device.[41] On average, House implantees eventually improved their hearing by some twenty decibels: enough to recognize a wide range of environmental sounds at least. So far as understanding speech was concerned, House only claimed that the device should aid substantially in lip-reading. He did not claim that implantees would be able to follow speech without lip-reading.

In October 1983, 3M applied to FDA for premarket approval (PMA) for the implant. They submitted data from 206 patients implanted with the earlier "Sigma" design (with its 15 millimeter electrode) and 163 with the short-electrode version, known as Model 7700. The FDA, as is its usual practice, ploughed through the pile of information provided and carried out its own tests on the safety of the materials implanted, the sensitivity of the device to temperature, physical shock, electromagnetic fields, and so on.

In November 1984, the House/3M device became the first cochlear implant to be approved for implantation in deaf adults (aged eighteen or over). The FDA ruled that the device was safe, although it noted that the long-term effects of electrical stimulation were unknown. They also ruled the device effective, although again with important caveats. Broadly speaking, the FDA concluded that the device "allows many implant patients to detect environmental sounds and conversational speech at comfortable listening levels," and that for some patients, the device "may aid" with lip-reading. No more than that. There was certainly no evidence that a deaf person would be able to understand speech without lip-reading. Moreover, 4 or 5 percent of Model

7700 users seemed to receive no benefit whatsoever. It was not known how these might have been identified in advance. The FDA required 3M to carry out a whole range of studies relating to safety, expected life, effectiveness in enhancing lip-reading, and so on after it had entered routine use. These are called "post-approval" studies. But the approval was vital, and its widespread coverage in newspapers across the country provided the implant with welcome publicity. President Reagan sent congratulatory letters to 3M and to William House. The implant thus achieved the high visibility that, earlier, had been both courted and feared.

Meanwhile, in Melbourne, Clark had found a more willing industrial partner through the Australian government's Department of Productivity. It was in the course of negotiations with the department that he came into contact with the second firm that figures importantly in this story. This was Nucleus, an Australian group of companies that specialized in medical electronics (including cardiac monitors and pacemakers). The Australian government agreed to provide financial support for collaboration between the University of Melbourne and Nucleus. Various technical improvements would have to be made: the size of the device would have to be reduced; a hermetically sealed biocompatible implant package would have to be designed, and a speech coding strategy and processor developed.[42] American collaborators would be needed. The American market is crucial to the success of a sophisticated and expensive medical technology, and to gain FDA approval, data from centers in the United States would have to be included in the submission. So in 1982, an American audiologist was hired with the job, initially, of reconnoitering the American scene. By 1983, collaborating centers in Iowa and Texas had been chosen. Other trial centers, in New York, Seattle, and New Orleans, soon followed. (The intention was to collaborate with centers in Europe too, but that was more problematic, and took longer to arrange.)[43] Their submission to FDA included data on eighty-six patients (thirty-six in Australia and thirty-six in the United States) ranging in age from eighteen to seventy-nine. One of them had been deaf for only five months, another for fifty-four years! As with the 3M device, FDA reviewed data on the performance of the device that had been submitted and carried out its own tests. Data showed, to the FDA's satisfaction, that the "Nucleus" device did offer enhanced communicative possibilities to its users. In 1985, the Nucleus implant became the second (and first multichannel system) to be granted FDA approval. The FDA's report went further than that regarding the 3M device because they saw this device as able to do more. The FDA concluded that not only could recipients detect sounds and distinguish speech from other kinds of sound, but for some patients it enabled them to follow speech without lip-reading, albeit to a limited extent.

Faced with the Nucleus challenge, 3M worked with the Hochmairs on redesigning their extracochlear "Vienna" device.[44] In 1985, with FDA approval of the multichannel Nucleus device imminent, 3M began to promote the extracochlear device as a worthy and less invasive alternative. However, this extracochlear device had not been approved, other than for limited experimental purposes, and could not be sold to clinicians interested in trying it out. In 1987, 3M decided to discontinue sales and promotion of the Vienna device. As attention in the field switched from safety to performance and as the idea that "more channels were better" gained ground, 3M had to acknowledge that it had fallen behind technologically. There was a third strand to 3M's strategy, however, and that entailed production of a multichannel device. After an approach to the Stanford group did not work out, 3M scientists set about developing their own device, known as "Sprint." Approval to start clinical trials was received in 1986, but given that a comparable FDA-approved device (from Nucleus) was already available, few people were willing to be implanted with this experimental device.[45]

At least three other small companies were also active in the United States in the early 1980s. They were Biostem, a small company collaborating with Blair Simmons at Stanford; Storz, a surgical instruments company that began working with the UCSF group after their collaboration with 3M broke down; and Symbion, a spin-off firm established by Kolff's group at the University of Utah. Some firms had committed themselves to a specific design or collaboration. Others, like 3M, had hedged their bets by investing in more than one version of the technology. But before there could be any commercial competition, the FDA regulators would have to be satisfied. William House and 3M were the first to cross this hurdle. Professional objections to use of the device were crumbling as professional bodies also endorsed the implant: the American Medical Association gave its endorsement in 1983 while the American Academy of Otolaryngology–Head and Neck Surgery followed suit in 1985.

Meanwhile in Europe, the French were making larger claims for the Bertin "Chorimac" device than the FDA would have accepted. Politics played an important role here. Thanks to the French Ministry of Research and Industry, the device had become linked to national innovation policy. Whatever its benefits for the deaf, the French implant was also seen as an opportunity for French technology to prove itself commercially. To enable it to do so, the import of competing foreign-made devices was discouraged. By 1978, Chouard had implanted forty-two patients including a ten-year-old deaf-born boy. "All of our forty-two patients have been able to hear," reported the Paris group. It is hard to be sure what they meant by this. Just three weeks after implantation, ten could apparently recognize two of ten "domestic sounds," and nine could identify three. By 1979, Chouard and his colleagues claimed that with good training and a sound-rich environment, it ought to be possible for speech to be understood without lip-reading. An implanted child must "leave the deaf school he previously attended, to be placed in a school for normal children, in an environment of normal speech and hearing." By the mid-1980s, groups in Grenoble and Lyon were also implanting the twelve-channel Chorimac, in close collaboration with Chouard. About sixty patients had been fitted with this large and cumbersome device.[46] But the data being collected in France would never satisfy the American regulators, and the French device would not be able to compete in markets outside France unless evidence for its effectiveness were based on some standard protocol. That evidence was gathered eventually. In the meantime, American colleagues were reluctant to take Chouard's results seriously.[47]

Despite their common faith in the technology, there was still much that the cochlear implant pioneers did not agree on. They continued to debate the relative merits of single-channel or multichannel designs, of intra- or extracochlear devices, of one coding strategy or another. This variety of conceptions of what precisely a new technology should be like is typical of the early history of many technologies. There is nothing unusual in the various pioneers—House, Clark, and the others—proclaiming the merits of his particular implant design. Cochlear implant pioneers were divided along the same "experimentalist" and "therapeutic" lines that Fox and Swazey had found among transplant surgeons. The experimentalists were still committed to grounding medical innovation in theoretical understanding: FDA approval was not the end of the road. The search for a theoretically optimal design had to continue. For their therapeutic colleagues, the only important test was whether the device worked. If patients could be helped, the search for theoretical understanding was of secondary importance and could come later. In the world of cochlear implant development, two decades ago, this division was clearly present. Associated with it was a competition for status and authority in the field. Ellis Douek recalled a conference that had taken place in the United States, at which William House had asked to be the first to speak:

> And he got up and he said, "Before we present our papers I think each one of us has to say first of all how many cases he has implanted." Because that would give us a clue as to the validity of their findings. Now everybody else crumbled, because he had implanted three hundred people, and everybody else had implanted twenty or something like that. But I was the second to speak. And I scotched it forever. I got up and I said, "I think that I have to explain to you first of all what we are doing, our team in London. I have to say that we are deeply grateful

> to Dr House because he has really made it possible for all of us to do the work. None of us would have any funding at all if it hadn't been for the publicity with which he surrounded . . . this is true . . . we are grateful. He also implants vast numbers of people . . . We don't do that. We are a research team, and we are paid by the Medical Research Council to do research not to treat people . . . Now if you have come here to buy an implant, buy Dr House's. He can sell it . . . But what we are doing is the *research* to give you the implant of tomorrow."[48]

As professional and commercial commitments became greater, as the first devices received professional and FDA approval and made their way to the market, the role of experimentalist came under pressure. Douek was in effect saying that a much better cochlear implant could be developed. This claim was more than a purely symbolic challenge. Some of his colleagues saw it as a threat to their professional interests. Trying to provide a clinical service using existing technology, they did not want a colleague around the corner saying, "Wait! A much better technology will be along soon." Blair Simmons's situation was even more difficult—as Douek saw it—because integration between medicine and industry had gone much further in California than it had in Britain.

With the gradual integration of scientific, professional, and commercial interests, the rules of the game, the sources of reputation and authority, were changing. The status of one design or another, one group or another, was no longer a matter of the quality of the science or the quality of the technology alone. Douek saw clearly that Blair Simmons and himself were being marginalized: excluded from the process through which industry and medical practice were aligning their interests. Nowadays, neither Douek nor Simmons occupies a place of honor in the professional retrospective. These places are reserved for the doctors whose commitment was to something practical, usable, and profitable. First, William House was the guest of honor at implant conferences. Then it was Graeme Clark. The threat from the experimentalists was beaten off.

Gradually, claims for the effectiveness of the implant and evidence in support of them were incorporated in publicity material designed to show physicians and patients what the implant could do. The growth of the market depended on convincing physicians and patients, as well as health insurers, that they could have confidence in the device. By the mid-1980s, it was clear that commercial expectations had been too optimistic. The market was growing far less rapidly than had been expected. Manufacturers viewed the slowly growing market with alarm. 3M launched a campaign to persuade physicians to promote the technology among patients. But still the results were disappointing. In 1985, 3M voluntarily recalled their 3M/House device from the market due to technical defects, following FDA product-recall guidelines. This recall was seen with anxiety even by potential competitors, as it threatened the image and the future of the technology. From 1986, 3M began to reduce its overall commitment to the field and then to phase out research on a more advanced design.

What lay behind the overall lack of market growth? Part of the explanation was financial. The costs of the device itself plus the surgery and hospital costs meant that cochlear implantation cost somewhere between nine and fourteen thousand dollars in the 1980s, depending on the device. Few deaf people would have been able to pay this themselves or (at that time) have had it covered by their health insurance plans. The future of the technology as a routine intervention would depend upon whether it would be covered by Medicare, for without such coverage the market would be unprofitably small. The nascent cochlear implant industry had a common interest here, even though commercial competition would later pit one manufacturer against the other. Under the auspices of the Health Industry Manufacturers Association, companies set about lobbying for Medicare coverage. They were successful. In September 1986, the Health Care Financing Administration (HCFA) issued a favorable coverage ruling for both single- and multichannel devices. But matters were more complex. As part of the set of policy initiatives in the mid-1970s designed to hold down the costs of health care, in 1983 Congress had passed legislation changing the way Medicare worked. Instead of reimbursing whatever "reasonable costs" were claimed, henceforth the HCFA would reimburse inpatient care in a "prospective" manner.[49] Mutually exclusive categories called "diagnostic related groups" (DRGs) were defined: 470 sets of pathologies. Average costs of providing treatment and other services for each DRG were established. Hospitals would be reimbursed within some range for each patient in a particular DRG. Although a mechanism was established to review advances in treatments that might merit an increase in the fixed payments for one DRG or another, clearly this system was going to have major consequences. Indeed that was the intention: assignment of a device to a high-paying or low-paying DRG was to be an instrument for regulating the diffusion of expensive technologies.

In the case of cochlear implants, the effects of this new system were profound.[50] Within the HCFA, considerable discussion took place over into which DRG cochlear implants would be placed, although not one of them would pay the full estimated cost of $14,000 for implantation of the (more expensive) multichannel device. So when cochlear implants were finally allocated to DRG 49—the most favorable from the hospitals' point of view—hospitals still faced a financial disincentive, to multichannel implantation at least. The result was that of the "170 hospitals involved in providing multichannel cochlear implantation, 10 percent have openly acknowledged to the manufacturer that they restrict the provision of the implant because of the loss of $3000 to $5000 that is incurred with each Medicare case." Medicare reimbursements were made for only sixty-nine cochlear implants in the 1987 fiscal year. This was one factor that led 3M to sell its ownership of the 3M/House device to the Nucleus-daughter Cochlear Corporation and then to leave the business completely. Nucleus/Cochlear did not share 3M's pessimism. Though the market was growing slowly, their own share of it was gratifying, and by the late 1980s, the Australian company's United States subsidiary had captured 90 percent of the American market.[51]

Something other than the costs of implantation was also involved in the slow growth of the market. Deaf people were not proving to be the enthusiastic consumers that they had been expected to be. Why not? Little was known of the motivations—or lack of motivations—of deaf people regarding implants. One industry representative speculated that perhaps they had become so accustomed to the world of deafness that they feared "entering the world of sound."[52] In some way or other, deaf people did not identify with the intended user of the device "inscribed" in the cochlear implant: with the behaviors, roles, and relationships ascribed to users. This failure of deaf people to identify with the technology provoked a strategic move on the part of the implant industry that was to change the rules of the game fundamentally.

It was becoming clear that the criteria by which implantees were selected had major implications for the size of the potential market. A study of hearing loss in the British population conducted in the mid-1980s showed how sensitive the potential market was to audiological criteria and age limits. If cochlear implantation was limited to postlingually deafened adults (eighteen to sixty-five years old) with a hearing loss of at least one hundred decibels, then the number of possible implantees was 12,056. If the hearing loss threshold was lowered to ninety decibels, then the number rose to 22,490. But if the prelingually deaf were added and the lower age limit dropped, then the number of potential implant candidates shot up to more than 64,000.[53] It is impossible to know what role these estimates played in people's thinking, and they were scarcely alluded to in public. Nevertheless, coincidentally or not, it was at this time that attention turned to implants for deaf children, a step that previously only Chouard in France had taken and that had drawn the opprobrium of his colleagues.

The Start of Pediatric Cochlear Implantation

At a conference on cochlear implants that took place in 1983, a panel discussion was devoted to the question of implanting children. The topic aroused a good deal of hostility. Dr. Claude Fugain, who had collaborated with Chouard for many years, explained the rationale for the step they had taken. Everyone agreed that hearing aids have to be fitted as quickly as possible after the diagnosis of hearing loss in order to maximize benefit, and the same reasoning applied here too. Experience had already suggested that the earlier the implant, the better the results. There were also the results of laboratory research on guinea pigs that suggested that early acoustic stimulation could prevent "auditory atrophy" in the brain stem. Not wholly spelled out by Fugain, but beginning to play a role in the discussion, was the notion of a critical period in neurological development. Curiously, neurophysiologists, whose field this is, were less inclined to take the notion of a critical period as proven. One of them, Gerald Loeb, intervened in the discussion:

> I just want to add a cautionary remark about this notion of the critical period, which we have suddenly decided is a real thing. First, the fact is that this is a very hypothetical notion. I don't know of any good evidence that it even exists in the auditory system. Second, the sort of pattern stimulation that we provide with electroneural prostheses may not fulfill its necessary inputs. Third, we have no idea at what age that period exists, or is critical, or ceases to exist. I think that it is worth thinking about, but I would certainly urge caution in taking this as some sort of blanket license to go around putting hearing aids or electroneural prostheses in very young children.[54]

Everyone knew that more was involved here than scientific evidence about pathways in the brain. How to respond to the demands of parents desperately seeking help, investing so much hope in new technology? Blair Simmons, another member of the panel, urged caution: "Parents have a lot of motives. All of which are usually to do good for their child, but not necessarily on a rational basis. I've seen this pertaining to oral education preferences as opposed to total communication. Some parents are unwilling to learn sign language. Others expect 'the doctor' to fix everything. Factors of this sort have to be under control because I don't think we'd ever want them to be the decision criteria for a cochlear implant." Touching on an issue that was later to become a central focus of controversy, a Dr. Owens from the audience referred to the possible value of sign language for deaf children. William House responded in terms that would be invoked for decades to come: "One of the greatest gifts that any of us can give our own children is our own language. Every parent does this with a normal hearing child. But if you cannot give your own child your own language, there is no way you are going to escape a tremendous emotional impact from that. Therefore, if the parent feels that they can give their deaf child oral language, then they should have every opportunity to try."

Few were in a hurry to follow Chouard and then House into the field of pediatric implantation. It was partly a matter of clinical caution, and partly a sense that the technology first had to be adapted before it could be used with children. Clark and Nucleus were developing a version of the device specially adapted for children. Clark implanted his first teenager, a fourteen-year-old boy, in 1985. Implantation of children younger than this would have to wait until the smaller device had been perfected. In 1986, the FDA decided to limit research on child implantation because of potential damage to the child's cochlea. In the course of that year, Clark in Australia implanted a ten-year-old boy with their new "Mini" device, and a few months after that a child of six. In 1987, in Los Angeles, William House implanted a Nucleus device into a five year, eight month old child.

It took somewhat longer for the implantation of deaf children to achieve general acceptance in the profession. In May 1988, the National Institutes of Health convened a "Consensus Development Conference" on cochlear implants. Following standard procedure, a day and a half of presentations by experts was followed by discussion. A consensus panel made up of specialists in medical and related scientific disciplines, clinical investigators, and public representatives weighed what they had heard. The panel's report reviewed many of the uncertainties surrounding the technology: criteria for implantation, benefits and risks, the advantages of the various types of device. The panel noted how variable the benefits seemed to be and how unpredictable in any individual case. They noted the lack of standardized tests, which made scientific assessment problematic. They noted the importance of a strong and interdisciplinary rehabilitation program. Further research on how auditory information is processed in the brain was needed, and methods for predicting individual success had to be developed. Children, the panel noted, do pose special problems. They grow, and their speech and language develop along with their other capacities. With children, it is often difficult to distinguish improvement due to a particular intervention from improvements due to general development. Specially designed studies using control groups were essential. And such studies had not been carried out: "No studies have adequately separated the effect of the implant from improvement due to maturation and training."[55] The conclusion was that although children could benefit from the implant, they must "still be regarded as hearing impaired" and will "continue to require educational, audiologic and speech and language support for long periods."

In 1989, a comprehensive review of progress in the implantation of children was published.[56] Dorcas Kessler, of the UCSF department of otolaryngology, summarized data relating to the two devices (3M and Nucleus) then approved for experimental use in children in the United States. French and

German clinical investigators, mostly using devices produced in France and Austria, often failed to separate results with children from those with adults, so that their work was of little relevance. As of mid-1987, 282 children in the United States had been implanted with the 3M/House device and as of mid-1988, eighty-three children with a Nucleus device. The tone of the review was optimistic but critical. So far as the House/3M device was concerned, publications provided no clear evidence regarding "the acquisition or maintenance of intelligible, spontaneous speech . . . the implant does not eliminate the need for special services and for consistent, ongoing training." And as for the extent to which the device facilitated lip-reading, no information was available. The children who did especially well were typically ones who had not been born deaf but had lost their hearing as the result of meningitis. There was much less information relating to children fitted with the Nucleus, although a few seemed to have done very well. Lack of standardized and longitudinal data was a major problem in reaching any definitive assessment. Dorcas Kessler concluded by noting, "Until such findings are available, it may be appropriate to decrease the rate at which young deaf children are being implanted, so that, when implanted, it is known that they will receive the best currently available system."

In 1990, a few months after this book appeared, the FDA approved use of the Nucleus implant in children aged two to seventeen years.[57] The approval was based on a clinical study of two hundred patients, implanted at twenty-five collaborating centers. The FDA had tried to establish "whether the device provided primary benefits of hearing sensation, environmental sound detection, and speech perception enhancement, and the secondary benefits of changes in speech production characteristics." Its conclusion was that "the cochlear implant gave a consistent improvement in the hearing of deaf children whose hearing was not adequately assisted by hearing aids. The children appeared to be a representative sample of the population under investigation. . . . [These] data . . . constitute valid scientific evidence adequate to provide reasonable assurance of the safety and effectiveness of the Nucleus Cochlear Implant when it is used in children ages 2 through 17 years."

In May 1995, the National Institutes of Health convened a second Consensus Development Conference to again review the state of knowledge regarding the technology. The data available were far more extensive than they had been at the conference in 1988. Certainly as far as adults were concerned, an unambiguous conclusion seemed justified: "Cochlear implantation improves communication ability in most adults with deafness and frequently leads to positive psychological and social benefits as well. The greatest benefits to date have occurred in postlingually deafened adults." So far as children were concerned, matters were less clear-cut. Cochlear implant outcomes were more variable in children. "Nonetheless," the conference concluded, "gradual, steady improvement in speech perception, speech production, and language does occur."[58] There were still major gaps in knowledge. That could hardly be denied. There was a troubling "unexplained variability" in the performance of implant users. In the absence of theoretically grounded longitudinal studies, little or nothing was known of the effects of implantation on children's language development. Despite these important gaps in knowledge, the consensus panel was convinced that the benefits of implantation in both adults and children had been established.

Intriguing questions arise. There was a large body of evidence for the effectiveness of the implant, but that evidence was limited in scope. Consensus had been reached on the basis of evidence relating almost exclusively to speech perception and production. Only minimal attention had been paid to the possible consequences of implantation for children's linguistic, cognitive, and psychosocial development. Why should this have been? And how could these many areas of ignorance, though acknowledged, have had so little consequence for the consensus process? The implant did seem to give deaf children better understanding of speech, and this is what everyone focused on. They did so partly because this is the variable with which hearing professionals (audiologists, speech therapists) work. They did so also because almost everyone concerned by the implant wanted it to work. Doctors, audiologists, teachers of the deaf, entrepreneurs, the parents of deaf children, and some among the deaf themselves all preferred to be convinced by what had been achieved than to worry about all that was still unknown. That could be filled in later.

A *Mature Technology?*

In the course of five decades, scientific doubts and professional opposition to the implant have been overcome or marginalized, criteria of eligibility continuously extended, and what once seemed an uncertain commercial future transformed. The device has become a commercial success. Today well over a hundred thousand people worldwide, about half of them children, are using cochlear implants.[59] Three manufacturers now dominate the world market, each of them offering a variety of models and each investing heavily in improving its technology. The Cochlear Corporation, claiming 70 percent of the market and reporting a $95 million profit in 2007, is the market leader. The other manufacturers are MedEl (an Austrian firm established by the Hochmairs in 1989) and California-based Advanced Bionics, established by medical entrepreneur Alfred E. Mann in 1993, that developed a commercial implant based on work carried out at UCSF.[60] Though the device remains extremely expensive, ear surgeons even in poor countries are enthusiastic about it. By 2005, nearly four thousand devices had been implanted in ten Latin American countries.[61] The cost of the implant has led researchers in a number of countries, including China, Korea, and India, to develop versions suited to their economic circumstances, though whether locally produced cheaper implants will have much effect on the market remains to be seen.[62]

As is typical with technological development, leading manufacturers came to dominate cochlear implant development. The shift from professional to corporate dominance of the process of technological change influences the directions in which development takes place. Alfred Mann suggests that innovation in the implant field might in the future be guided by a new paradigm. Instead of viewing cochlear implants as a means of "treating deafness," and thus as an end in themselves and subject only to further incremental improvement, Mann explains that he began to see them as belonging to an emerging *class* of neuroprosthetic devices. In a recent interview, he explained this creative leap in his thinking:

> In Advanced Bionics we were producing a cochlear implant, and the business was moving along very nicely, but I felt that for the company to remain strong and independent, it needed to diversify. And so we began to look at other places where we could use our technology a little differently. We reasoned that a cochlear implant is, after all, the most sophisticated of neurostimulators. So we looked for neural markets that we felt were underserved, and we decided that spinal cord stimulation could be an interesting opportunity to pursue. This was a market dominated by one company, where the technology was not meeting market needs. And it was an area where we felt that if we could come up with disruptive technology, we could change the marketplace.[63]

It is impossible as yet to say what such reconceptualization of the implant—no longer as the "superior sort of hearing aid" that it has sometimes been made out to be, but as the first of an emerging class of neuroprosthetic devices—might mean. That it could have significant implications, for the future of the technology and for the professionals having jurisdiction over it, seems probable. One can be sure, however, that if it is successful, such a reconceptualization will locate the implant even more firmly in the realm of the "bionic future" that *Scientific American* celebrated.

Chapter 3

The Cochlear Implant and the Deaf Community

For a while, in the mid-1980s, cochlear implantation seemed to be faltering. Despite endorsement by the medical profession, the market was growing far more slowly than manufacturers had expected. It was recognized that financial barriers were holding back sales of the device and that hospitals faced financial disincentives. Nevertheless, the discovery that few adult deaf people seemed even to want an implant was unexpected. Neither manufacturers nor implant teams had much idea of what lay behind this lack of interest, though it clearly involved something more than just the costs. Implant teams and manufacturers then turned their attention to deaf children, the vast majority of whom have hearing parents. Smaller versions of the implant that allowed for their wearer's growth were developed and new rehabilitation procedures established. In 1990, market prospects were dramatically transformed by FDA approval of the Nucleus implant for use with children.

Why were adult deaf people so unenthusiastic, despite professional endorsement of the implant and despite the widely publicized claim that it could give them hearing? As discussed in chapter 1, there are various grounds on which people reject interventions (such as prenatal testing and vaccinations for their children) recommended by medical practitioners. Some people consult the medical literature themselves and find the evidence for the intervention insufficiently convincing. Others reason on the basis of deeply held personal convictions or beliefs, or the prior experience of themselves, friends, or family. Refusal of a medically prescribed test or treatment does not necessarily mean doing nothing. Beliefs or experiences that lead a person to go against a physician's advice may also lead him or her to seek out an unconventional form of treatment or (self) care. It may lead him or her to seek out other like-minded people, looking for support from others who have made the same difficult decision. A variety of health-related practices and biosocial groupings derive from critique (and perhaps rejection) of some part at least of medical orthodoxy.

One is the diverse set of healing practices commonly called "complementary and alternative medicine," or CAM. The orthodox medical view has long been that holistic approaches to healing were mere quackery. At best they were seen as offering solace and comfort. Nevertheless, it is now clear that CAM is used to an increasing extent, especially by people suffering from life-threatening diseases such as cancer.[1] Its techniques and practitioners give people something they feel is missing from orthodox medicine. Today, American consumers spend billions of dollars annually on CAM.[2] Another group who reject traditional medical treatments is the environmental breast cancer movement, which calls the biomedical model of the disease into question but then moves in a more political direction. Criticizing the emphasis on individual bodily etiology, it argues that attention should shift to the environmental causes of the disease.[3] Another example is the "pro-ana" movement, which brings together, rather loosely, people who reject the orthodox view of anorexia as an eating disorder.[4] Pro-ana material is disseminated over the Internet through a wide variety of support groups and social networking sites. At some of these sites, anorexics can discuss their problems, the strategies they use to diet or hide their weight loss, or their experience in refusing medical or psychological treatment. Some sites go further, insisting that anorexia be viewed not as an illness but as a "lifestyle choice" that should be respected by families and by doctors.[5]

The medical profession has responded differently to the different approaches. However reluctantly, CAM has to some degree been embraced. A National Center for Complementary and Alternative Medicine was established under the NIH in the 1990s and many leading academic medical centers have established departments or institutes for the study of CAM. In the United Kingdom, some National Health Service hospitals have begun to provide CAMs for cancer patients who want to make use of them.[6] Without engaging the epistemological differences involved, and through careful processes of demarcation and adjustment, clinicians and medical institutions are finding ways of using CAMs to complement their professional role and strengthen their market position.[7] The medical response to the pro-ana movement has been very different. Rejecting established approaches to managing (that is, curing) anorexia, pro-ana has been vigorously attacked by the National Eating Disorders Association, by the National Association of Anorexia Nervosa and Associated Disorders, and in the media. In both France and the United Kingdom, bills were introduced into national parliaments that sought to make pro-ana sites illegal. The French Minister of Health, supporting such a bill, stated that "giving young girls advice about how to lie to their doctors, telling them what kinds of foods are easiest to vomit, encouraging them to torture themselves whenever they take any kind of food is not part of liberty of expression."[8]

CAM poses a fundamental challenge to medical thinking. Its practitioners question the methods and epistemology underlying randomized clinical trials, the gold standard of medical proof. Riuping Fan, a Hong Kong–based scholar who has studied the relationships of traditional Chinese medicine to western biomedicine, lists their objections.[9] In traditional Chinese medicine, according to Fan, emotional states and experiences of patients are taken to be objective facts, as open to study as biochemical or physiological indicators. Disease processes are unique to the individual and need to be studied through individual-sensitive observation-based studies, not through grouping patients on the basis of diagnostic categories as is done in randomized trials. But CAM does not challenge medical consumerism. Through providing CAM services for the growing number of patients who want them, medical professionals and institutions have discovered that they can enhance their market position. By contrast, pro-ana perspectives offer no market opportunities and pose a different kind of challenge. In enabling adherents to share their "thinspirations," routines, and regimens, to know that they are not alone, the pro-ana movement provides its adherents with a form of empowerment that is far removed from that propagated within the medical model. In other words, pro-ana views seem not only to threaten medical authority but at the same time undermine the way in which patient empowerment is defined in current medical discourse.

What then of deaf people and their response to the cochlear implant? The first response, and the one that clinicians and manufacturers noted with surprise and alarm, was indifference. Market forecasts that had anticipated a significant proportion of deaf people in the United States seeking an implant had been wrong. *Outsiders in a Hearing World, a* book published in 1980 that few doctors or industrialists are likely to have read, might have shown them how they had gotten it wrong. The author, Paul Higgins, was the child of deaf parents. His wife taught deaf children. He himself had spent a year teaching in a school for the deaf before taking up graduate studies in sociology. In the introduction to his book, Higgins explained how it had grown out of an attempt to make sense of his personal experiences. "Making sense" involved locating those experiences in a broader social context: thereby "transforming personal experiences into sociological issues."[10] Subtitled "A Sociology of Deafness," the book painted a radically different picture of deafness from the clinical one. Though excluded from a hearing world in which they nevertheless live, the deaf are not the social isolates they were commonly held to be. Higgins portrayed a complex community with which many of its members strongly identify. Deaf clubs provide their members with a place for easy social intercourse. Like home, and for some perhaps more so, the deaf club provides a refuge from the grinding frustrations of the hearing world. Here they can communicate freely and easily with their friends. The basis of that communication is sign language. "Signing is not a sufficient condition, though it is a necessary condition, for membership in deaf communities," Higgins wrote. "Signing is an indication of one's identity as a deaf person and one's commitment to the deaf world. It is perhaps the most obvious indication to hearing people that one is deaf."[11] Higgins's analysis shows why few deaf people were interested in the implant. The expectation that large numbers of deaf people would be eager for an implant was based on a lack of understanding of their lives. Though the hearing world could be unwelcoming, even hostile, and the work environment could be lonely, deaf people have their own community. At home with their families, among friends, in the deaf club, deaf people enjoy social lives little different from those of their hearing neighbors. The key to all this is sign language. Deaf people who sign, and who are integrated into the Deaf community,[12] do not see themselves as in need of a prosthesis to help them speak. Deaf people's lack of interest in the implant was due to the fact that most of them did not see their lives as unfulfilling or their bodies as defective in the way that doctors and manufacturers assumed.

A New Deaf Consciousness

That deaf people's response to the cochlear implant subsequently went beyond this initial indifference has to be understood in relation to the history of sign language, the basis of their sociality.

For hundreds of years, the education of the few deaf children fortunate enough to receive any education was based on the attempt to teach them to speak. The story has often been told of an eighteenth-century French priest, the Abbé de l'Epée, and how his encounter with two young deaf women led him to see the importance of a gestural mode of communication in teaching the deaf. Since deaf people could communicate freely in such a mode among themselves, de l'Epée reasoned, a teacher could best use it in teaching. His efforts led to the founding of public schools for deaf children throughout France, including in 1794 the National Institute for Deaf Mutes in Paris, in which signed language was the medium of instruction. In 1814, de l'Epée's deaf assistant Laurent Clerc traveled to the United States and introduced the new sign language–based education there. This was the history that Harlan Lane recounted in *When the Mind Hears*, the book so reviled by David Wright.[13] But then, at the end of the nineteenth century, opinion among educators of the deaf turned against using sign language. School after school reverted to the older oral tradition, and parents were discouraged from letting their deaf children communicate using sign language. This argument was based not on science but on ideologies. Historian Douglas Baynton explains it in terms of a transformation in the motives behind educating the deaf.[14] For clergymen like de l'Epée, the goal had been to give deaf people access to scripture. The vital thing had been to establish the most effective means of achieving this end. But ideas about participation in society then changed. Not moral development but work-related skills, participation in the workforce, came to be seen as crucial. Emphasis on speech at the expense of sign language–based education reflected this shift in what participation in the community meant. To work with hearing peers, to take one's place in the economic order, implied an ability to use the language of the workplace. So sign language, the deaf person's route to personal and moral development, was replaced by total devotion to teaching deaf people speech, and through it the practical skills, the useful arts, with which they could enter the workforce. The result was that starting in the late 1880s, signing deaf teachers vanished from schools for the deaf and sign language–based education disappeared.

The autobiographical stories of older deaf people, who grew up in the earlier part of the twentieth century, testify to how the dominant hearing culture viewed sign language and its users. They tell of being taken from their parents at an early age and sent to a residential school for the deaf. They tell of a harsh regime in which endless hours were devoted to arduous and often fruitless attempts at teaching them to speak. They tell of their first encounter with signing children, and the wondrous new possibility of easy communication. But sign language was typically banned in the classroom and sometimes even in the playground or dormitory. Teachers forced children to sit on their hands when they were seen signing, or beat them on the hands. This education did not provide them with much. Few deaf school graduates could easily read a newspaper, let alone write a letter.

Despite the horrors of the classroom, these deaf children came to value their signing peer group to the extent that, for many, returning to the loving but often communication-less world of the home became not a matter of longing but of regret. This is a very different image from familiar cultural representations of boarding school, of a tearful little boy (it is usually a boy) left behind with his trunk of belongings: there to be made a man but crying himself to sleep every night. How could this be so different? How could it be that the affective bonds that deaf children developed with each other could become more positive, more rewarding, than the nurturing environment of home? It becomes comprehensible when one knows how little contact many of these children had with their families. Growing up deaf in the 1940s and 1950s was often lonely and embittering.

Older deaf people all have such stories to tell. Their stories are a catalogue of "experience after experience of the frustration of being denied sign language and of the harshness of those who forbade it."[15] The frustrations of growing up in a world that largely barred them from easy communication formed part of the common culture of deaf people. Their own social life, by contrast, centered around the deaf club, provided relief, companionship, and community. There, among friends, experiences could be shared, stories told. There it did not matter what the hearing world thought about sign language. Unlike in the outside world, deaf people did not have to feel ashamed of communicating in sign.

In 1957, just before Charles Eyries performed the world's first cochlear implant operation, William Stokoe joined the faculty of Gallaudet College (now University) in Washington, DC. Gallaudet is a unique higher educational institution for deaf and hearing-impaired students. Stokoe had written his PhD dissertation on medieval romance at Cornell University, and he was appointed a professor of English. On joining Gallaudet, Stokoe "began to learn how to produce signs, which were then presented to us as equivalents of specific English words."[16] Gradually he began to doubt that it was quite that simple. Not a fluent signer at the time, he nevertheless became interested in the possibility of applying tools of formal linguistic analysis to the signing he saw about him. Encouraged by linguists at nearby Georgetown University, he began work. Many people at Gallaudet, including the deaf students, thought the project ridiculous. But Stokoe persevered with the help of deaf assistants who signed for him before a camera. In 1960, he published a monograph on the structure of American Sign Language (ASL). The analysis showed that deaf people's signs had all of the formal properties of a full language. Comparative analysis came soon after. In 1961, Stokoe went on to study sign language in Britain, and gradually linguists there and in other countries followed his lead. Presenting his work to Gallaudet colleagues, Stokoe later explained, did not evoke much enthusiasm in these early days. Not only was the Deaf community skeptical, colleagues in deaf education—"as hostile to Sign Language as ignorant of linguistics"—responded to his work with icy disdain.[17]

Stokoe was not to be deterred. In 1965, he and his colleagues published a dictionary of American Sign Language; in 1970, Gallaudet established the Linguistics Research Laboratory under his directorship; and in 1972 a new journal, *Sign Language Studies,* was started. Nor were linguists the only scholars becoming intrigued by sign languages. At the world-famous Salk Institute in California, Ursula Bellugi and Edward Klima were studying how the brain processes language. Because it was so very different, a visual language like ASL offered the opportunity of adding to neurobiological understanding of language in general.[18]

By the late 1970s, experts in the nature of language, neurophysiologists and linguists, were convinced that sign languages were real full-fledged languages. As more and more signs were recorded, and as more and more of the grammar of sign languages was analyzed, it became possible to teach sign languages like any other languages. But all this research was not enough to overturn a lifetime of prejudice. It was still common for the hearing world to dismiss the manual communication of deaf people as "a mishmash of pantomime and iconic signals."[19] Deaf people knew better. They had used sign language for generations, in their everyday lives, in intellectual discussion, in poetry and drama. "Yet even there the hostility of the dominant culture sometimes led the deaf themselves to incorporate the hearing world's assessment of their language as a primitive pidgin, a gestural analogue of 'You Tarzan, me Jane.'"

The 1971 Congress of the World Federation of the Deaf, held in Paris, caused a stir in Europe. European participants were astonished to hear that American educators of deaf children were stepping back from the complete oralism of earlier years. There was new enthusiasm for the approach to teaching known as "total communication." The important thing now was to communicate with the children. Precisely how was of less consequence. Increasing use was being made of English accompanied by signs, or "sign-supported English." This combination of speech and signing seemed to make for effective communication. Perhaps signing did have a place in the schools for the deaf? However, there were many in Europe who saw any reintroduction of signs as a retrograde step, not to be countenanced. The following congress, held in Washington, DC, in 1975, caused an even bigger stir among European participants. As far as the French delegates were concerned, it strengthened a feeling that the congress in Paris had hinted at. French Sign Language (Langue des Signes Française, LSF) could have a status like that which, it seemed, American Sign Language was gradually gaining.

By late 1975, attitudes were beginning to change in France. A few articles on sign language appeared in periodicals. A weekly television program for deaf and hard of hearing people began in which sign language was used. The French National Association of the Deaf (Confédération Nationale des Sourds de France) set up a committee to study problems of communication, including the idea of "total communication." In November 1976, Bernard Mottez, a sociologist attached to the Center for the Study of Social Movements (Centre d'Etude des Mouvements Sociaux) in Paris, and Harry Markowicz, a sociolinguist from Gallaudet, began a remarkable research project. They wanted to record the process of emancipation of deaf people in France that they were convinced had begun, and they wanted to participate in and contribute to the process. Mottez and Markowicz started a newsletter, *Coup d'Oeil*, as a public forum for the exchange of information. In the first issue, which appeared in January 1977, a new seminar series was announced: Sign Language and the Deaf Community, to take place at the prestigious School of Advanced Studies in the Social Sciences (EHESS) in Paris. Through a series of seminars, they introduced the work of the major American sign linguists. Thanks to the work of Bill Moody, an American sign language interpreter living in Paris who interpreted into French sign language, deaf people could participate. This was the first seminar series at a French university institution to be interpreted into sign language. At the same time, the first "half clandestine" courses in French sign language had started.

Reporting in 1979, Mottez and Markowicz wrote of their hope "that the participants in the seminar would begin to make waves around them. They did, because they represented the whole scale of involvement in deafness: from deaf people and parents of deaf children to professionals." People participating in the seminar helped establish the country's first bilingual (French/LSF) kindergarten, a biweekly TV program in which fairy stories were told in LSF, and began to plan an Academie de la Langue des Signes Française.[20]

In 1980, Bernard Tevoort, professor of linguistics at the University of Amsterdam and another sign language research pioneer, gathered information from around Europe on the status of sign languages.[21] Awareness of sign language was on the increase in almost all countries. And although there were certainly national bastions of oralism, the general tendency in schools for the deaf seemed to be in the direction of total communication. Oralism was starting to give way, and in one way or another, signs were reentering the world of deaf education. Tervoort saw this as the beginning of a longer-term shift that he thought would lead ultimately to the use of true sign language as medium of instruction.

The work that Stokoe and his colleagues had started at Gallaudet was spreading outwards, attracting interest far beyond the community of professional linguists. As Bernard Mottez and Harry Markowicz had hoped, it was stimulating people to press the claims of sign languages socially and politically. It was stimulating cultural and organizational initiatives of many kinds. The late 1970s were years of growing ferment, of growing deaf consciousness. It is necessary to understand this to make any sense of what the work of Stokoe and his colleagues meant to deaf people. For a deaf person, the Gallaudet environment at the time could be a source of transformation, of self-discovery, of a new understanding of what it meant to be deaf.[22] If the gestural communication that deaf people preferred to use was no primitive pantomime but a true language, why be ashamed of using it? Why not take pride in its remarkable expressive qualities? Out of this research grew a pride in being deaf that led to the emergence of sign language–based art forms such as the National Theater of the Deaf, and to new political demands.

The changes taking place were illustrated powerfully and elegantly by two deaf scholars, Carol Padden and Tom Humphries, in a book that appeared in 1988 in which they tell of a visit to France.[23] Many of the deaf people they met were anxious to recount the story of the Abée de l'Epée to them. Why so much emphasis on this one individual in all these historical retellings, Padden and Humphries asked themselves? There were plenty of other things that could have been recounted. Why was there so little interest in telling of the founding of the local deaf club, or the story of Laurent Clerc, who in 1814 left France for America to help found a school for deaf children there? "We finally realized that the story is not about the Abée de l'Epée. Instead it has come to symbolize, in its retelling through the centuries, the transition from a world in which deaf people live alone or in small isolated communities to a world in which they have a rich community and language. This is not merely a historical tale, but also a folktale about the origin of a people and their language."[24] It is the symbolic significance attached to this episode that makes it so important.

Historians may find that the origins of sign-based instruction in France were more complex than this, or that de l'Epée changed his sign-grammar to correspond to that of spoken and written French, that it was not in fact French Sign Language. That is not the point. The power of the folktale remains. Padden and Humphries go on to describe their viewing of a film made in 1913 of a lecture by George Veditz, then the president of the (American) National Association of the Deaf. Veditz's film was made as part of a project initiated by the National Association of the Deaf and designed to preserve sign language. An historical retelling, but also a call to arms, the lecture was delivered on the thirty-third anniversary of an event that also has great symbolic importance in deaf people's historiography. Veditz told his audience of the 1880 World Conference for the Deaf in Milan, at which hearing educators of the deaf agreed to abandon sign language–based education and return to the use of speech. Padden and Humphries evoke the vivid images with which Veditz, in 1913, urged his audience to fight for the retention of sign language.[25]

Events like these, the life and work of the Abée de l'Epée and the tragedy of the 1880 Milan Congress, have a great importance for deaf people, and opportunities for recounting the stories are much valued. The basic theme of the historiography constructed around them is one of enlightenment and the flourishing of a deaf culture followed by the onset of a dark age. The tales invoke hope and the possibility of redemption. This history is meant to work for the deaf just as medical professionals have a preferred history that works for them. It must give meaning to achievements, and it must provide consolation in the face of adversity. But Padden and Humphries's own widely acclaimed book, the work of two deaf university teachers, has a different significance. Its publication by a renowned university press, its scholarly rendering of such central themes in deaf culture, also announced that the dark age was ending. All this was going on in parallel with—but totally disconnected from—development of the cochlear implant. It was this new sociohistorical and political consciousness that provided a basis for a perspective on the cochlear implant that was radically different from how much of the medical profession viewed the device.

The Implant in the Mass Media

Over the course of the 1980s, the implant was gradually accepted in medical and industrial circles. As clinical data accumulated, clinicians became enthusiastic, manufacturing corporations interested, regulatory bodies convinced, and health insurers willing to reimburse implantation. A network of professionals and implantees emerged for whom the value of the technology was beyond doubt. The information circulating in this network, all the articles attesting to what could be achieved with the implant, reinforced their commitment and trust.

In order to understand how the implant came to be understood by deaf people and how a view different from mere indifference emerged, one has to look at similar factors. Here too the circulation of information was crucial. But the information that circulated among deaf people was different from that to which clinicians had access. Accounts of the cochlear implant that deaf people encountered were not statistical reports in international medical journals but reports in the popular press and on television that they read, saw, or were told about by friends and acquaintances.

Study after study of the development of medical technologies has emphasized how the mass media systematically reinforce popular faith in what seems to be the limitless potential of modern scientific medicine. Each new wonder technology attests to the validity of the imagery and the reasonableness of the hopes. Organ transplantation, for example, seems to offer double redemption: to the donor family, which now has the opportunity of turning tragic loss into the gift of life; and to the recipient, whose hopes of life are rekindled. The transplant surgeon's heroic work makes this possible. It is these themes—redemption, hope, the heroism of the surgeon—that the mass media like to emphasize.[26] "Findings from research involving cell cultures and laboratory animals become harbingers of imminent human cures," writes lawyer and ethicist Rebecca Dresser.[27] She shows how the "almost promiscuous use of the term *breakthrough*" reflects both career interests of scientists and journalistic incentives to sensationalize.[28] So it was in the case of the cochlear implant. Media stories stressed the possibility of "making the deaf hear." Press reports painted a much more positive and unqualified picture of what cochlear implants could do than could be found in the medical and scientific literature. And though these reports were local, and so differed from place to place in the events they recounted, the responses of Deaf communities all drew on their newly emergent political consciousness.

Some of the early work on cochlear implants attracted little or no media attention. Douek's work, for example, was conducted within a strictly research framework and was reported only at scientific conferences and in journals of medicine and audiology. It received little or no public attention. Recruitment of the handful of volunteers needed for research could take place via existing contacts, and there was no need for publicity. But some of the other pioneers were rapidly convinced that, even in its relatively primitive stage of development, the device could be used to help deaf people. Publicity could be crucial in attracting the attention of potential implant candidates who would otherwise be hard to reach. The French surgeon Claude-Henri Chouard, for example, soon realized that publicity would be important for the continuation of his work. In a memoir published in 1978, Chouard is explicit about it. He realized that people become enthusiastic about "the magic words: discovery, science, hope. . . . Television served us well. In the course of a broadcast that took place in March 1977, undecided potential patients and parents were given an objective picture of what we were doing and of its current limits; of our doubts and of our hopes. The serious and careful tone of these few minutes were convincing, because they showed the essence of our work: that is, they showed—live—word comprehension."[29]

Reports in the newspapers and on television had two strategic purposes so far as Chouard and some of the other implant teams were concerned. Media reports were intended to impress funding committees—and they did. Though all the publicity did not endear Chouard to his professional colleagues, it did succeed in attracting the funding he needed. Publicity was intended also to attract the attention and awaken the interest of potential implant candidates among the deaf. And indeed, just as Chouard had hoped, among those who learned of his work were a few people who saw themselves as potential candidates for implantation and approached him. But as he moved on to implant deaf children, the first surgeon in the world to take this step, his work evoked another, very different, response.

In France, a group of articulate middle-class parents of deaf children had banded together in 1965 to form an organization, the Association Nationale de Parents d'Enfants Déficients Auditifs (ANPEDA). These parents were angry. They were dissatisfied with the provision made for their children's education. They did not approve of their children being segregated in special schools. What ANPEDA members hoped to achieve was integration of their children in regular education, and they were not going to let vested interests and established traditions stand in their way. ANPEDA soon became a highly effective organization, backed by substantial resources and enjoying close links with the medical and audiological professions. Committed as it was to oral (mainstream) education of deaf children, ANPEDA was initially enthusiastic about the possibilities of cochlear implantation. In 1973, ANPEDA, the Fédération des Sourds de France, and a number of organizations of teachers and social workers founded UNISDA (National Union for the Social Integration of the Hearing Impaired) to speak on their collective behalf, at least on issues on which they agreed. In April 1977, UNISDA issued a statement in response to the publicity that Chouard's work was attracting.

This statement, which appeared in the periodicals of a number of these organizations, stressed the need for caution and drew attention to the dangers of exaggerated publicity. Only a small number among the deaf were likely to profit, at least for the moment. Little was known of the physiological effects of implants. The risks and uncertainties in implanting children were considerable. All in all, the signatories insisted "while recognizing the good intentions and the seriousness of the research, [we] nevertheless invite all those involved to show extreme prudence. The confidence of deaf people and of their families risks being destroyed if the results of the operation fail to correspond with the accounts that are being given. The confidence of deaf people and of their families risks being destroyed by hasty communications not based on controllable results."[30]

At first opposed only to the exaggerated publicity Chouard's work attracted, ANPEDA was then outraged at what it saw as the technique's premature application to children. Whatever their hopes for the longer term, they were horrified at the idea of deaf children being used (as they saw it) as guinea pigs in experimentation. ANPEDA managed to secure support for their views from a leading figure in French otology. Michel Portmann was the son of an eminent ENT surgeon, grandson of the founder of the specialty in France, and a member of the French senate. Professor Portmann's status led to his objections achieving national publicity. Thus in February 1979, the daily newspaper *L'Aurore* reported, "Professor Portmann is indignant: 'The artificial ear is premature and dangerous.'" Portmann had taken advantage of a press conference held by his research institute to make his dissatisfaction with the state of affairs clear. The cochlear implant may well have had promise, and it certainly needed to be studied, but scientifically and discretely, not in the glare of publicity. Given the state of development of the technology, claiming to offer something to deaf people, and above all to deaf children, was totally unacceptable, Portmann told his audience. "It's scandalous. To spread by a book, by interviews resounding in the mass media, the hope that the deaf—the deafened and even the born-deaf—can now hear, is unacceptable."[31]

Only twenty-two people had been implanted in France, and there was no published data of an acceptable scientific quality about the results. Publicity was out of all proportion to the reality. Not only was the device extremely expensive (seventy thousand francs, of which only twenty thousand would be reimbursed by Social Security—implying a considerable financial sacrifice for the recipient or recipient's parents), but given how little was known of its working or optimum use, it could even be dangerous. Portmann cited the example of a little girl totally deaf in one ear but with some hearing in the other. She had received an implant in the better ear with the result that she had lost the little hearing she had. Other French ear surgeons shared some of the views expressed by Portmann. They too thought the technique promising but that its use with adult deaf should be more thoroughly explored before children were implanted. And they too deplored the publicity that the device had attracted. But few were willing to be publicly critical of a colleague.

A few years later, in Britain, the organization of parents of deaf children (the National Deaf Children's Society, or NDCS) reacted similarly. Their outrage attracted media attention too. By the early 1980s, a group of British ENT surgeons had emerged who did not share Douek's interest in developing the technology. They were keen to provide a clinical service, as House was doing. It was with this push towards clinical application that publicity began to develop. In 1984, the procedure was brought to the attention of the general public. The heroine of the story, whose photograph adorned the daily newspapers, was Jessica Rees. In 1984, Rees was twenty-one and still studying at Oxford University. In August of that year, Andrew Morrison, an ear surgeon at the London Hospital, fitted her with a cochlear implant. Rees was subsequently seen on television, humming along to music. Journalists' attention was drawn to the event, reporting, for example, "Hope of restoring girl's hearing with electrodes implant," and continuing, "she said: 'It is hard to explain but I am going to have to learn how to hear again. I will be hearing sounds that I have not heard before and I will have to learn what they are.'"[32] Later the same newspaper reported, "Electrodes work for deaf girl. Jessica Rees, 21, the deaf Oxford undergraduate who had a pioneering operation at the London Hospital on Tuesday, last night heard her first sounds for 17 years."[33]

Following the operation on Rees, NDCS became as concerned as ANPEDA had been, largely because of the flood of inquiries to which the publicity had given rise. The NDCS director, Harry Cayton, explained their point of view: "Medically these operations are still experimental. Five or six operations on adults in Britain do not yet provide a basis for extending the program to children. Nor do we know what the effect of having the implant for 10, 20 or 30 years is going to be. It might be very interesting for doctors and audiologists to study the reactions of a child with an implant but sometimes they seem to forget that deaf children are people, not just a set of non-functioning ears."[34]

As in France earlier, the objections of the parents' organization attracted the attention of the media. Yet reporting of the NDCS position was not universally sympathetic. Where one newspaper reported that Jessica's treatment was "'raising false hopes.' Surgeons warned on ear operation,"[35] another portrayed the organization as standing in the way of progress. Under the headline "Charity Blocks 'Bionic Hearing,'" this article in the London *Sunday Times* of November 12, 1984, continued, "A new 'bionic' ear implant that could bring hearing and speech to totally deaf children is being blocked by one of the country's leading charities for the deaf."

The newspaper's use of the term "bionic ear" is significant. It evokes a quite different image from the term "cochlear implant." For Carlo Laurenzi, a sociologist working for the British National Deaf Children's Society, the term "bionic ear" invoked the limitless potential of science. It is resonant with hope: the hope of a triumph over deafness.[36] This was not accidental. The term had been coined in Australia with precisely these associations in mind in the course of planning a televised fund-raising appeal for the Melbourne implant program. Someone had apparently suggested that it was as though they were building an ear for *Six Million Dollar Man*, a popular television program at the time.[37] Though there were some doubts about a term that hinted less at corrected hearing than at super hearing, it nevertheless proved irresistible. Melbourne soon became home to the "Bionic Ear Institute."

The term "bionic ear" appealed both to implant surgeons looking to mobilize resources and the mass media looking to reflect popular faith in medical science.[38] In both Britain and France, organizations of parents of deaf children were unhappy with this kind of reporting.[39] Since, in their view, it was far from clear to what extent deaf children could profit from the implant, it was necessary to proceed with caution. They were worried that publicity around the "bionic ear" would evoke unrealistic hopes and expectations. The concerns of

ANPEDA and NDCS focused principally on uncertainty, on the inadequacy of available knowledge. Although the parents' organizations were able to gain attention for their anxieties, their cry of "caution" received a not altogether favorable reception from the mass media.

The parents' organizations, ANPEDA and NDCS, did not represent deaf people in France or in Britain. Their leaders were middle-class hearing parents, many of whom had no contact with signing deaf people and no particular sympathy for sign language use. Moreover, the notion of a Deaf *community*, increasingly figuring in the rhetoric of the deaf, was not a reassuring one for hearing parents at that time. People knowing little of deafness, or encountering it for the first time, try to relate it to other more familiar chronic conditions, like asthma or rheumatism. Or they put their hands over their ears and try to imagine what it must be like to be deaf. From there, it is a large step to the idea that deaf people are tied together by social and cultural bonds, by a shared history of oppression and a shared language. A leading figure in ANPEDA, writing in the association's journal *Communiquer* in 1982, was critical of the concept, arguing that in the French situation it could only be divisive: "Alas, in the last few years, the concept of 'deaf community' has undergone, in France, a series of manipulations that result in the non-signing being banished to the distant shadows. The psychological obstacles that can separate a man from his fellow sufferer have been as it were codified, exploited, to oppose one to the other. This phenomenon has its origins in the USA. No one can any longer ignore the changes to which it has given rise in that country over the past decade."[40] As members of the Deaf community began to formulate their own response to the publicity surrounding cochlear implantation, the concerns they expressed were different from those of the parents' organizations, with the notion of community playing a central role.

The Bionic Ear and the Deaf Community

Initial indifference to the cochlear implant reflected a way of life that, for those who were part of the Deaf community, was not the isolated existence deaf people were popularly assumed to lead. The outside world knew little of this community or of deaf people's views of the implant, and few deaf people had the literary or social skills or the self-confidence to approach the mass media. In the 1970s, this began to change. The situation in France at that time is particularly fascinating. On the one hand, Claude-Henri Chouard had been successfully cultivating the mass media; on the other, *Coup d'Oeil* had been launched and emissaries from Gallaudet enthusiastically welcomed. For anyone committed to the furtherance of sign language and Deaf culture, anyone who had become involved with the new organizations and cultural institutions (including a flourishing deaf theater group), the media reports of the implant were a provocation. The timing was significant. Few in France had yet forgotten the political ferment of 1968. People remembered how close that theatrical outpouring of intellectual frustration, combined with worker organization, had come to bringing down the established order. And by the late 1970s, there were intellectuals working with the deaf who certainly did not lack the skills or self-confidence to make themselves heard. In December 1977, a group of French deaf people prepared a text that was subsequently publicized by a writer, Jean Grémion, associated with the Paris-based deaf theater group IVT.[41] It expressed a concern over the cochlear implant very different from what UNISDA had had to say somewhat earlier:

> We deaf, what do we see in all the newspapers and on the television: "Extraordinary invention of Dr. Chouard—17,000 deaf mutes can hear and speak." We smile. Why not bleach the blacks and blacken the whites? When are they going to stop, once and for all, using us as guinea pigs? We are astonished that everyone is talking about this invention, while normally there's never a word about our life and our world. For years there hasn't been a single line about us in the press . . . when are they finally going to accept our world as a reality? . . . Many among us are married, have children, and these children are happy . . . It is society that has oppressed us and continues to oppress us. The proof: this invention of Dr. Chouard. We speak only with a gestural language, that is our maternal language, and it is marvelous to speak that language.

The view expressed here would have been much less readily understood by most people than that of the UNISDA. The need to proceed cautiously in introducing a new medical technology seems reasonable. Grémion's text evokes a different theme, and one with which few people outside the Deaf community were familiar in the late 1970s. Embodying as it does the new self-consciousness of deaf people, equating being deaf with being black, it suggests a very different perspective on deafness than its identification with hearing loss.

Through the late 1970s and early 1980s, the cochlear implant received a good deal of attention in the periodicals of the French deaf organizations (*Communiquer*, the journal of ANPEDA, *La Caravelle*, that of the late-deafened, and *La Voix du Sourd*, the periodical of the French Deaf Association). But there was a growing divergence in perspectives. The journals of two associations (that of the parents of deaf children and that of people who had become deaf in adult life, who are rarely fluent in sign language) expressed a growing interest in the implant. However, *La Voix du Sourd*, the journal of the signing deaf, offered its readers little more than the occasional snippet of information.

The situation in France is illustrative, but the critical rhetoric around implantation that emerged was not a specifically French achievement. Deaf advocates and (hearing) researchers in various countries took exception to widely reported claims that went far beyond what the evidence showed and were based on assumptions that had little or no foundation in medical expertise. In developing their critiques, they drew on the new understanding of deaf people as forming a community united around the use of sign language. Additionally, they assessed the evidence for the benefits of implantation against medicine's own avowed standards of proof. Whatever its symbolic significance, the implant was also a medical device, subject to the same rules of evidence as other medical devices. Its technical utility had to be proven with hard statistics if health care resources were to be made available. The claims made for the implantation of deaf children in particular were critically assessed, and found wanting, in each of these respects. Critique, in turn, provoked a response from clinicians and oralist educators of the deaf who were convinced of the value of the procedure.

Weighing the Claims against the Evidence

In early 1991, Heather Mohay, a medical psychologist at the University of Queensland, published an article on deafness in children in the *Medical Journal of Australia*. She pointed out how poor had been the educational results achieved by oral methods and how, subsequently, attitudes to the use of manual communication became less rigid. Although few (Australian) schools were willing to introduce sign language and few deaf children had any contact with native signers in their early years, there was growing support for at least some use of sign (typically in the form of signed English).[42] Mohay said this was better than nothing, but only one step on what remained a long road. Then she turned to the promise of cochlear implants: an "impressive technological breakthrough" but no panacea.

> Both the British Deaf Association and the National Deaf Children's Society in Britain have made policy statements cautioning against the expectation of a "miracle cure." The statement from the National Deaf Children's Society concludes: "The Society is committed to a whole child approach where education, health, social and personal needs are dealt with together. Cochlear implants must only and always be part of that approach and not an alternative to it." Many deaf adults place a

> high value on their language and their culture and prefer to see themselves as a cultural minority group rather than as disabled. As a result many of them are opposed to the use of cochlear implants.[43]

However measured and reasonable Mohay's discussion seems, it was read as a rejection of pediatric implantation. In a letter to the editor of the *Journal*, the principal of a school of hearing-impaired children stated, "Children . . . have the right to the language of their family and since 95% of all hearing-impaired children are born into hearing families, they have the right to spoken English."[44] The director of an implant program wrote an article-length response in which he attributed opposition to the implant to a mixture of ignorance of the results and self-interest.[45] His article displayed many of the prejudices and assumptions to which deaf people took exception: "Sadly a child deafened between two and five years may forget all speech and behave as if born deaf"; "The fact that most of the [implanted] children can use speech as their primary means of communication and need only learn to sign if they or their parents wish is a historic breakthrough"; "Some deaf people prefer Australian sign language or Auslan, which can be used to communicate, and to express ideas and concepts. It is a valid language although it lacks the grammatical structure of spoken and written English"; "The problem remains that of the deaf children. Should they become part of this deaf culture group or should they join the hearing world?"

As a scientist, Mohay was doing no more than asking for the evidence on which medical interventions are supposed to be based, displaying a proper scientific skepticism. The responses her article evoked, however, referred not to evidence but to the "right" of children to the language of their parents (taking for granted that that language was English) and assumed a lesser status for sign language on the ground that its grammar was different from that of English.

In a book appearing in 1992, Harlan Lane, author of *When the Mind Hears*, provided a comprehensive review of the evidence, particularly that for pediatric implantation.[46] His review was very different from that published by Dorcas Kessler just three years previously. Lane looked carefully at the variety of tests that had been conducted on implanted children, tests with names like GASP and PBK. Instead of asking the standard question—"Do the children do better with their implants than they did without them?"—Lane asked, "How well do they actually do?" His interpretation of the literature was that the different tests gave significantly different answers, but that on average the children could not actually identify much that was said to them without lip-reading. Lane criticized authors for indiscriminately pooling results from children who had been deaf from birth with children who had been deafened after learning

English. If a study group contained patients who had became deaf later in life, their average scores would be higher than for a group of only deaf-born subjects. In this way, argued Lane, a false impression was being created. He tried to look at the findings from the point of view of a questioning parent. "The first and fundamental question that parents have for the doctor is: Can you make my deaf child hear? They generally do not mean: Will implant surgery give him any hearing at all? They mean, rather: Will he be able to hear well enough to learn our language, to communicate with us, with his teachers, and with other hearing people? In short, they want the deafness undone." So far as Lane was concerned, the clinician, reviewing the research literature, could say no more than that such an outcome was statistically unlikely. Little could be said regarding any particular child's likely benefit.

Lane's review of the cochlear implant literature emphasized the enormous gaps in knowledge of the effects of implantation. In his reading of the literature, there was no evidence that an implanted child benefitted sufficiently to pass through school as a hearing child. At best, he argued, the deaf child could function as a severely hard-of-hearing child. And so far as school achievement was concerned, severely hard-of-hearing children did little better than deaf children. These are the kinds of things that parents are concerned with, or at least should be concerned with. In deploying the skills of a psychologist and a statistician to show that data are (or were) not very good at all, Lane's critique turns medical orthodoxy on itself.

Also in 1992, the first European Symposium on Pediatric Cochlear Implantation took place at the University of Nottingham, in the United Kingdom. One of the keynote speeches was delivered by a leading professor of otolaryngology from the Albert Einstein College of Medicine in New York. Robert Ruben stressed the vital importance, to the developing child, of developing a competence in language use as early as possible. He stressed that a distinction had to be made between speech and language. Language is not the same as speech. Medical science had identified all kinds of situations in which normal speech accompanies profound language deficits. What was crucial in the care of the deaf child was optimizing language acquisition during the critical period of the child's development. Ruben pointed out that almost all the literature on outcomes of cochlear implantation in children focused on speech production and reception. Almost none of it looked at language. In the present state of the technology, Ruben argued, the cochlear implant could not provide a sufficient "flux" of aural language input during the vital second and third years of the child's life. Only sign language could do that. The implication was that the cochlear implant had to be used to augment the input of visually based language to which the child did have adequate access. "It is a disservice to the child," Professor Ruben told his distinctly uncomfortable audience, "to obtain better speech at the expense of language." His audience of ENT surgeons, audiologists, speech therapists, and others engaged in cochlear implantation was not happy with this lecture. It was not what they had expected to hear from a leading member of their profession. They were not pleased to be told that they did not understand what language was. They did not want to be asked to think about sign language.

A body of literature that seemed to Kessler, as well as to regulatory and professional bodies, to provide grounds for cautious optimism seemed at best inadequate and at worst misleading to critics. The many ENT surgeons and audiologists who were by now convinced of and committed to implantation did not welcome this skeptical questioning of the evidence, whatever the supposed importance of skepticism in scientific medicine (and indeed science). It is important to understand where these two divergent interpretations of the literature come from.

Pioneers like Chouard, Clark, and House were convinced that it was becoming possible to alleviate deafness and its consequences. They were doing their best to help desperate people coming to them. That is how they interpreted their experiences, experiences that had been important to them as physicians. They had to be willing, as they saw it, to take above average risks in the hope of developing new ways of helping. They saw themselves standing firm in the face of colleagues' conservatism and doubts. They drew on the testimony of some of their grateful (though sometimes disappointed) patients who had been called to bear witness to the value of their implants. It was in these terms that the central actors in the implant's development recounted their achievements. The themes evoked, the commitment in the face of adversity, the motives and passions, the ultimate triumph, have become the standard repertoire of medical innovators. Challenges are confronted and overcome.

One physiologist ridiculed the status patient testimony had been given, saying the value of a new procedure cannot be demonstrated with letters from patients. "Hard data" are required, argued Nelson Kiang, and an implant can only be designed on the basis of deep understanding. Not so, was the rejoinder. Medicine has always progressed despite the doubters. Let's see what works in practice and find out why it works later. Kiang's rebuke to William House in 1973 would hardly be necessary today. As resources for health care became tighter, clinicians were obliged to accept that they needed hard evidence: they needed large numbers of patients and randomized clinical trials. In private, doctors, like patients, are often persuaded by witnessing a single remarkable cure. But publicly, that will not suffice. Those who control the purse strings will only be convinced by statistics. As resources became tighter and new treatments competed with each other for funding, the rules of the game changed. Numbers became all-important.[47]

Nevertheless, the positive reading of the cochlear implant literature rests ultimately on faith: faith in the benefits of medical advance and faith in its unbounded future possibilities. Medicine's history, in which the cochlear implant is inscribed, shows the progressive accumulation of more and better ways of helping sick people. The development of new tests, techniques, and therapies is a hard road. There are conservative colleagues unwilling to support necessary research. There are the skinflints of the insurance industry unwilling to pay. There are intellectuals raising ethical doubts from the comfort of their book-lined studies. But sustained by faith in the possibilities of medical science and by professional commitment, the intrepid pioneer presses on, determined to ease the pain and suffering of patients. Questioning the evidence is unwelcome because it betokens a lack of faith.

The alternative view, as expressed by Lane and others, also rests on values and beliefs that are not reducible to scientific evidence or to standards for assessing the quality of evidence.

The Rights and Wrongs of Cochlear Implantation

Deaf people could hardly fail to be struck by the contrast between the enthusiastic popular and media response to the implant on the one hand and the lack of attention and respect for their way of life on the other. The exchange between Mohay and her critics provoked a further response from two member of the Centre for Deafness Studies and Research at Australia's Griffith University. They tried to explain that deaf people were not asking for very much: "Members of the Deaf community wish to be seen as ordinary people who communicate in sign and other modes. They seek a greater understanding by the hearing community of their status as another realization of the human condition. They ask that the social, educational, vocational and personal implications of this status be understood by the hearing community, especially by those advising parents of young deaf children."[48] More was at stake than their wish for acceptance as "ordinary people who communicate in sign." Deaf people also felt that cochlear implantation threatened the Deaf community as a whole. A short time before this exchange appeared in the *Medical Journal of Australia*, the Australian Association of the Deaf published an "Open Letter to Deaf Australians." In it they wrote, "Someone said that implants (cochlea or bionic) should be developed and implanted in every profoundly/severely deaf child and then those children when they grow into adulthood can be assimilated into the hearing world too . . . And then there will be no need for deaf committees, deaf centers and deaf clubs in 40–50 years' time. *Whoopee*!!! . . . What does that mean? What do you as a *Deaf Australian* think of that? . . . If you, like me, think this idea is ghastly, we should dig our heels in and stand up for our rights." Elsewhere others were reaching similar conclusions.

Paddy Ladd, founder of the radical National Union of the Deaf, and today a major figure in the field of deaf studies, was one. In the *British Deaf News*, he too reflected on the significance of the device, placing it in an historical context. His article began with a brief characterization of attempts, from 1880 onwards, at making deaf children hearing and at suppressing deaf culture.[49] The cochlear implant, for Ladd, represented the "fourth phase" in this terrible history: "The medical profession, for so long the enemies of the deaf community, are eager to get their hands on deaf children of our community. We will make these children hearing, they imply, if we can't get oralism to work any other way, then we'll make them into hearing children; as if they could." The Deaf community, Ladd argued, has a responsibility towards deaf children and must mobilize to protect and preserve their rights to grow up as members of the Deaf community. He called on members of the British Deaf Association to protest at a forthcoming international conference on deaf education that had picked implants as its major theme. His article ended with the unforgettable slogan "Cochlear implants—Oralism's Final Solution."

In explaining the benefits of the implant, the mass media—but also clinicians—continually appealed to all the cultural images and metaphors to which deaf people took exception. The way the implant was presented by the mass media, on the basis of individual success stories, made it all the worse. One example of this piling metaphor on metaphor was the response to a broadcast of the American television program *60 Minutes*, televised in 1992. It told the story of a captivating little girl called Caitlin whose life had been transformed by a cochlear implant. This child's story captured the enraged attention of activists in the Deaf community. "The mildest criticism was that Caitlin's success was a fluke that would tempt parents into entertaining similar but doomed hopes for their own children. 'There should have been parades all across America,' Caitlin's father lamented months later. 'This is a miracle of biblical proportions, making the deaf hear. But we keep hearing what a terrible thing this is, how it's like Zyklon B, how it has to be stopped.'"[50]

Harlan Lane developed the most thoroughgoing and influential critique of the cochlear implant.[51] In addition to criticizing the quality of the evidence for its effectiveness, Lane set the device into an historical context. The context he invoked was not that of medical progress, inspired by faith in the unlimitedly benevolent possibilities of medical science. It was not the context that provided cultural legitimation for new medical technologies in general and for the "bionic ear" in particular. Lane's history of the cochlear implant related it to the suppression of sign language and deaf culture and to the succession of "heroic" attempts at curing deafness.[52] Lane rooted his critique in themes that have become familiar in social scientific writings. All about us the standing, the self-image, and self-interest of the medical profession are mobilized in the attempt to reformulate social problems in terms of individual pathology. It is politically expedient to consider criminality or drug abuse as medical problems, subject to treatment and—once gene therapy has gone a little further—to prevention. A drug or a prenatal test can be used to turn deviance from a social norm into a correctable or avoidable pathology. With time, the alternative—to accept short people, fat people, disabled people, or deaf people as simply different—loses its legitimacy. Invoking Michel Foucault, Lane presented the cochlear implant as one of the many "technologies of normalization" that modern medicine has provided. Do we want to live in a world that insists on a uniform standard of performance, health and beauty for all, and punishes or treats those who deviate from the norm? Or would we rather live in a world that tolerates—and indeed rejoices in—physical and cultural diversity? For Lane, the answer was clear. What is more, the right of cultural and linguistic minorities to protection is enshrined in a United Nations declaration: the Declaration of the Rights of Persons Belonging to National or Ethnic, Religious, and Linguistic Minorities. "The Declaration calls on states to foster their linguistic minorities and ensure that children and adults have adequate opportunities to learn the minority language. It further affirms the rights of such minorities to enjoy their culture and language and participate in decisions on the national level that affect them. Programs that substantially diminish minority cultures are engaged in ethnocide."[53]

Deaf opposition to implantation was also formulated in bioethical terms. This is not really surprising. Bioethics had initially been conceived as providing a means of debating the implications of advances in biomedicine for human welfare and for the human condition. Its fundamental principles, including "respect for persons" and "justice," seemed well-suited to the concerns that Harlan Lane and others were trying to formulate. Moreover, bioethics has achieved a high degree of legitimacy and, backed by bodies like the World Health Organization, claims universal applicability. A convincing bioethical case against pediatric implantation would be a powerful weapon. That medical professionals working with the implant sought to refute the arguments made is thus hardly surprising.

Translation of the controversy into the language of ethics focused on three areas of disagreement. One concerned deaf children's development.

Implant programs in many countries insisted (as many still insist) that parents commit themselves to the exclusive use of spoken language. Through emphasis on the use of speech at home and attendance at an oral school the value of the implant is maximized. The child is enabled to participate in mainstream hearing society. Critics argued that this is damaging to the child. Such practice neglects what from the child's point of view is a more valuable option: to grow up as a member of the Deaf community.[54] Not only are the results of cochlear implantation limited, certainly when measured in terms of functional participation in the hearing world, but membership in the Deaf community offers the child the vital benefit of a positive sense of self. Oral education had failed, and to bring children up in a context that emphasized and sought to overcome their disability was no service to them. Since membership in the Deaf community was psychologically and socially so much more beneficial, argued ethicist Robert Crouch, even were the device much more effective than it actually was, this participation would still be more desirable. Harlan Lane and Gallaudet professor of deaf studies Benjamin Bahan pointed to another danger. Anticipating that their child would receive an implant, parents of children newly diagnosed as deaf would not bother to learn any signs. Thus, however much or however little the child might ultimately profit from the implant, it would be deprived of access to language for a vital period of its early life.[55]

Related to this was another theme: the social and cultural status of deaf children. Lane and Bahan argued that the child "in the normal course of things" would become a member of the Deaf community and so have values different from his or her parents. Deaf children "have a DEAF heritage from birth" as a consequence of their physical constitution. Their cultural status is determined by "the culture the child would enter in the normal course of things." Noel Cohen, a well-known New York cochlear implant surgeon, disagreed: "Deaf children of hearing parents are not members of the deaf community until they are either placed in that community by their parents or voluntarily decide to enter it." For him and for his colleagues (and for many parents), implantation was seen as offering the child the possibility of later choosing whether or not to enter the Deaf community. As far as Lane and Bahan were concerned, what was at stake was whether or not children be allowed to take possession of their birthright.[56]

The third, and most important, claim drew on a central theme of medical ethics, and had to do with clinicians' responsibilities. To whom or what is the doctor responsible? Clinicians typically stress that their principal responsibility is to the individual patient seeking their help. As Cohen put it, "We . . . have an obligation to serve our patients to the utmost of our ability. Since cochlear implants have become available for the treatment of the profoundly deaf, and since these devices have gained wide-spread use and have been approved by responsible government agencies in many countries, . . . it is hard to see how we can be accused of unethical conduct."[57] Critics argued that there are overriding considerations, and that limits must be placed on the clinician's freedom to act in the interests of the individual patient. For Cohen, if pediatric implantation has deleterious effects on the Deaf community, this is regrettable but not something that can be allowed to interfere with his responsibility as a surgeon. For some critics, these effects were more important than what the doctor wanted to do for the individual patient. Lane and Bahan stressed that arguments based on the notion of deafness as disability, which might be used to justify medical intervention, were inapplicable because the deaf have a distinctive culture while disabled people are acculturated to mainstream values.[58] Like Crouch, Lane and Bahan posed this ethical problem in an extreme form by assuming a 100 percent effective implant. If used on a large scale, this would lead to a virtual elimination of Deaf culture: they call it "ethnocide." Such a consequence would not only be against the interests of the Deaf community but, as they put it, would impoverish the world's cultural diversity. "Difference and diversity . . . are a major part of what gives life its richness and meaning." That this is a greater good than that of the individual patient is to them self-evident.

The problem in making this bioethical critique stick is one that bioethical scholars themselves have raised and that was discussed in chapter 1: namely, the primacy that principlist ethics places on the individual patient's right to informed and autonomous choice. It is in terms of these principles that medical professionals today largely understand their ethical responsibilities. In this respect, Cohen was expressing a view that most of his medical colleagues would have shared. The notion of autonomous choice is of great value to physicians or patient groups insisting on a right of access to any procedure or technology.[59] A powerful resource, it has been widely used in justifying a variety of controversial interventions. "Each woman should have the right to decide," for example, was an argument plastic surgeons used in their campaign on behalf of silicone breast implants, despite the risks that had been identified.[60] Appeal to the patient's right to autonomous choice fails to acknowledge the tension between a doctor's responsibility to his (or her) individual patient and the obligations of the medical profession to society as a whole. While some bioethical scholars recognize that this exclusive emphasis on fidelity to the individual patient, omitting any notion of a broader social and ethical purpose, is problematic, their views have had little impact on most physicians' understanding of their responsibilities.

In the early 1990s, both the (American) National Association of the Deaf (NAD) and the World Federation of the Deaf (WFD) adopted much of Lane's analysis.

> New and high technology that entails invasive surgery and tissue destruction is used, not for life saving, but for putative life enhancement. Adults, such as these children will become, when given the option of such prostheses, overwhelmingly decline them. The parents who make the decision for the child are often poorly informed about the deaf community, its rich heritage and promising futures, including communication modes available to deaf people and their families. Far more serious is the ethical issue raised through decisions to undertake invasive surgery upon defenseless children when the long-term physical, emotional and social impacts on children . . . have not been scientifically established.[61]

By turning the cochlear implant into the "bionic ear," the mass media embedded it in an account of medical progress that went far beyond the detail of technical achievement. Medical technologies are just as much a symbol of progress as, for example, space exploration. Both evoke human power over the unknown. The way the mass media write about these technologies draws on very basic ideas, or metaphors, that we use to make sense of life and to celebrate life. The power of medical technology is as much to do with its being a symbol of hope as with its technical effects.[62] To refer to the implant as the bionic ear was to draw attention away from its technical effects and towards its symbolic ones. It was no longer audiograms that were evoked, but the Six Million Dollar Man with his superhuman capabilities. As the claim that the "bionic ear" offered a "cure for deafness" gained currency, Deaf community leaders came to see the device as symbolizing all that stood in the way of their emancipation and their aspirations.

In response to the claims made for the implant, between the 1970s and the 1990s a critical discourse that had two major elements was crafted. One deployed the tools of science itself in order to criticize the way in which professional consensus was reached. It can be seen as a continuation of the arguments put forward by the investigators called "experimentalists" in the previous chapter. Roughly speaking, it expressed a conviction that closure was occurring prematurely. The second, and more central, element had to do with the lives that deaf people led and the meaning of deafness for them. Two aspects of deaf experience, as they understood it, underpinned the discourse critical of the cochlear implant. One was a new sense of deaf history: a history now seen as marked by a century of rejection of sign language and deaf culture. The other was an invigorated sense of community. Members of the Deaf community shared an experience of exclusion and stigmatization and they shared a common language: sign language. Though there are parallels with the demands of disability rights advocates, in that both denied the relevance or the appropriateness of a medical model, the deaf emphasized their shared language and culture. The Deaf community, for those socialized into it, was a source of identity, and was becoming a source of pride too. This new self-confidence, this new pride in sign language and deaf culture, led to demands for their acceptance in the education of deaf children, for sign language interpreters, and for other societal accommodations. Coincidentally or not, these demands took shape in precisely the years in which the medical profession and its industrial allies were developing the cochlear implant. It is this commitment to the right of the Deaf community and Deaf culture to protection, respect, and a degree of self-determination that is the essence of Lane's position.[63] It is this right that he saw pediatric cochlear implantation as violating.

In 1993, a sympathetic journalist writing in the *Atlantic Monthly* referred to a survey in which 86 percent of deaf adults said they would not want a cochlear implant even if it were free. He went on to quote Harlan Lane: "'There are many prostheses from eyeglasses and artificial limbs to cochlear implants,' Lane writes. 'Can you name another that we insist on for children in flagrant disregard of the advice of adults with the same 'condition'?"[64] Comparable with the response to the pro-ana movement, and in striking contrast with the rhetoric of "experience-based knowledge," the experience of deaf people living with their deafness was not accepted as relevant by the medical profession. Nevertheless, as the technology spread around the globe, these opposed understandings of implantation played out in ways that differed from country to country. How and why that was so is the subject of the following chapter.

Chapter 4

The Globalization of a Controversial Technology

DESPITE THE CONTROVERSY, the cochlear implant is now a commercial success. Growth has been considerable and is expected to continue. Two decades ago, less than two thousand people had been implanted worldwide. Today a single manufacturer reports that its 120,000th device has been implanted, and that in 2007 nearly 16,000 devices were sold.[1] Despite the controversy, the cochlear implant is now a global technology, too, having spread from rich countries to middle-income countries, and now to the poorest of countries. The leading manufacturer, Cochlear, is present in no less than ninety countries. That is to say, in the course of approximately two decades, cochlear implant operations have been carried out in nearly half of the countries of the world. Over the same period, the implications of implantation have been the subject of continuing discussion within the World Federation of the Deaf, for which the rights of deaf people and the status of sign languages remain issues of the highest priority.

In order to understand the globalization of the technology and of the debates and practices associated with it, it is necessary to focus on two sets of issues.[2] The first is "flows": to "trace movements of people, things, images, ideas from one local world to another. This means not only recognizing the flow of phenomena but specifying . . . the channels through which they move and the landscape changes such flows might bring about." One central claim of this chapter will be that the channels through which a medical technology flows are very different from those through which a discourse contesting that technology flows. Specifically, promotion of the cochlear implant and the critical discourse of deaf community advocates spread in rather different ways. The second issue concerns what happens at the point of arrival. That is to say, one must examine the ways in which "the global" is conceived, or invoked, in specific local contexts and projects. For example, how is reference to the United Nations Declaration of Minority Rights put to work in a specific national political or legal dispute? How precisely a new medical technology is deployed is shaped by its integration into a particular local world, with its specific values and its specific political economy of health. Many factors play a role, including the extent to which a particular health care system is centralized or decentralized and State run or market driven, the availability of resources, and the extent of specialization of the medical profession.[3]

Genetic testing, for example, has been shown to be organized in ways that reflect patterns of funding and organization of health care in one country or another.[4] These features of health care systems influence who has access to the technology, and how. A second central claim of this chapter will thus be that wherever one looks, the opportunities for deployment of the technology on the one hand and the critical discourse of the deaf on the other also differ significantly one from the other.

Arjun Appadurai, a professor of anthropology at the University of Chicago and leading theorist of globalization, has emphasized the importance of these differences to understanding globalization processes.

> The various flows we see—of objects, persons, images, and discourses—are not coeval, convergent, isomorphic, or spatially consistent. They are in what I have elsewhere called relations of disjuncture. By this I mean that the paths or vectors taken by these kinds of things have different speeds, axes, points of origin and termination, and varied relationships to institutional structures in different regions, nations, or societies. Further, these disjunctures themselves precipitate various kinds of problems and frictions in different local situations. Indeed, it is the disjunctures between the various vectors characterizing this world-in-motion that produce fundamental problems of livelihood, equity, suffering, justice, and governance.[5]

How implant technology has become global, despite the objections and the arguments of the Deaf community, can be understood in terms of the disjunctures to which Appadurai refers and their distinctive consequences in each national situation. The argument is elaborated by first examining the beginnings of implantation practices in three countries, and the ways in which the arguments of deaf activists and advocates were (or were not) brought to bear. On the basis of these descriptions, it will be possible to elaborate on the chapter's two central claims, namely, that globalization of this contested technology can be understood in terms of differences in flows or pathways on the one hand and in opportunities for putting technology and critique to use at the national level on the other.

Policies and Practices

The Cochlear Corporation envisaged an international market for the technology from the start. In 1984, just two years after the parent company had been founded in Australia, Cochlear established a subsidiary company in the United States. Three years later, a second, to serve the European market, was established in Switzerland. Growth in the United States was much more rapid than in Europe. By December 1988, more than a thousand Nucleus devices had been implanted in the United States. Compared to this, there had been 179 in Australia (where the manufacturer was of course based), and in the United Kingdom, Sweden, and the Netherlands, three, two, and one respectively. Since some implant teams had chosen to use devices produced by other manufacturers, these figures do not correspond to the total number of implant operations carried out in those countries.[6] They do, however, give some idea of the relative scale of activity. What is more, establishment of those programs, however small, was the result of considerable work on the part of the surgeons who set them up.

UNITED KINGDOM

In Britain, the presence of Ellis Douek and his colleagues had a major influence on the initial course of events. Supported by the Medical Research Council (MRC), Douek was committed to the development of an implant that would be superior to anything then available. His work, and its support by MRC, was an implicit statement that available technology was not yet good enough. Not everyone shared this view. By the early 1980s, there were a number of doctors who wanted to provide a clinical service.

One of the first to act was Graham Fraser, a senior ear surgeon at London's University College Hospital. Fraser did his first implant operation in March 1982 using a prototype implant then being developed at UCSF.[7] Fraser tried to obtain funding for his work from both the Medical Research Council and from the government Department of Health and Social Security (DHSS). Both turned him down. The MRC was already supporting Douek, while the DHSS viewed the procedure as experimental and thus not (yet) eligible for National Health Service funding. Was the unwillingness of MRC or DHSS to fund his work due to opposition from the Deaf community? "Not at all," Fraser explained, "except that they will pick on anything as an excuse for not giving money."[8]

Working "on a shoestring," Fraser moved ahead. A little later, however, he was able to obtain some funding from a charitable organization that would see him through the following five years. Dissatisfied with the performance of the UCSF device, Fraser next tried out the device being developed by the

Hochmair team in Vienna.[9] Disappointed with this, too, Fraser decided that he would develop his own implant in collaboration with the Royal National Institute for the Deaf.[10] But this work would have different objectives from those of the other programs, including Douek's. Fraser wanted to develop the simplest possible implant: a single-channel device that, by virtue of low cost, could be made widely available.

Fraser's difficulties in finding the resources he needed and the search for private funding that followed set the scene for the expansion of cochlear implantation in Britain. Mark Haggard, the well-informed director of the MRC's hearing research institute, noted what was going on with some alarm. In 1986, "cautiously optimistic" regarding the technology, Haggard was nevertheless worried by the claims being made for the implant. It was being discussed as though the major problems lay in the areas of technological advance and surgical practice. Not so, argued Haggard. The much more complex issues lay in the areas of selection and rehabilitation. Because it was treated as an essentially surgical procedure, cochlear implantation was becoming disconnected from other forms of rehabilitation for deaf people. That was wrong. Implantation was simply one among the battery of rehabilitative procedures available. Drawing attention and funds from other practices and procedures, it could even work against the interests of deaf people in general.

In an insightful analysis, Haggard outlined some of the forces acting on the field of cochlear implantation in Britain. The first was the dominant role played by surgeons, partly responsible for the growing (and in his view unfortunate) disconnection from other forms of provision for deaf people. The second was industrial interest, or "entrepreneurial push." British efforts "lack the thrust" of those in the United States or Australia precisely because of "lack of any entrepreneurial push from a large British company that foresaw profits from sales of the hardware." This could of course be an advantage. A real concern for Haggard was the publicity being given to implantation by the mass media. He shared the concern expressed by the director of the National Deaf Children's Society, Harry Cayton.[11] Cayton had been concerned by the fact that the "balance of acceptance and hope on the part of thousands of parents of deaf children" was being disturbed by publicity. "That delicate balance," he went on, "is a commodity upon which the adjustment of deaf children crucially depends."

This was not the kind of thing anyone wanted to hear. In the absence of any financial commitment from government sources, surgeons in city after city turned to charitable bodies and to their local communities for resources needed to establish an implant program. Addenbrookes Hospital in Cambridge started in 1987, the Royal Infirmary Manchester in 1988, four other centers in 1989, another five in 1990. As NDCS staff member and sociologist Carlo Laurenzi observed, this need to raise funds from within the community reinforced the tendency to make inflated claims for the device, the practice that worried Haggard. At the same time, by continuing to treat the whole area solely as a research field, Douek—according to his own interpretation as well—posed a threat to the credibility of their claims that an established practice already existed. Douek was convinced that this led to a deliberate attempt to marginalize his work. The younger surgeons, as he saw it, "just wanted to buy Nucleus things, which were being sold . . . Nucleus company was putting in a *tremendous* sales pressure. They were setting up offices with *experts*, so that they could say 'Look, we'll show you how to do it, we'll send you on a course, we'll pay for you, we will test your patients . . . we will really play nanny to you . . . just buy it.'"[12]

The risk to this new generation of implant surgeons was that the DHSS could use Douek's work as proof of the experimental status of the procedure, and thus as reason not to provide funding. "So . . . they had to defend themselves by saying . . . 'That group is . . . rubbish. They haven't done anything of interest for the past three or four years . . . It's a dead duck. The rest of the world is implanting these things. It is *not* a research project. It *is* an available prosthesis, like a hearing aid.'"

In 1989, the Council of the British Association of Otolarynologists issued a statement that in their view, cochlear implants had moved from the stage of experiment to an established therapy for the deafened. By this time, some sixty people had been implanted in Britain, all using private funds. Government funding was seen as vital if further work was to be done. Previous approaches to the Department of Health and Social Security had been fruitless. At this time, with Graham Fraser in the lead, the British Cochlear Implant Group was established, with the explicit objective of lobbying for government support. Jack Ashley, a deafened Member of Parliament (now Lord Ashley) and chair of the House of Commons Committee on Disablement, arranged a meeting with a government minister at the DHSS. The meeting was carefully planned. "Most important of all we took a patient along. Because my psychology was that what had convinced me was to see a patient in Los Angeles."[13] The strategy paid off. In 1990, the government announced that £3 million would be made available over a three-year period for a National Cochlear Implant Program. Six centers would be selected for funding on the basis of competitive tendering. Comparable programs were established in Scotland and in Northern Ireland a little later.

Shortly before this, in 1987, the implantation of children had begun in Britain. When it did, the NDCS was as concerned as it had been at the earlier publicity about Jessica Rees. Ten years after Chouard had taken this step in France, an eight-year-old boy became the first British child to be fitted with the single-channel device developed by Fraser and his UCH/RNID team. Events then moved rapidly. In 1988, a specialized pediatric program was established in Nottingham to fit children with multichannel intracochlear implants. The NDCS was still wary and anxious to avoid a public controversy that could only confuse and worry parents. Gradually, however, they came to accept the value of a "carefully controlled, properly funded and cautious cochlear implant program for carefully selected children."[14] By 1990, the NDCS, more inclined to compromise than was its French sister organization, was engaged in difficult discussions both with the British Cochlear Implant Group and with the British Deaf Association (BDA). There were limits to how far clinicians were willing to go, while the BDA, though generally opposed, still had no official policy.

In Britain, the Deaf community had difficulty in formulating its position on the implantation of deaf children. Radical and conservative members of the British Deaf Association finally agreed a policy statement in 1994 after long delay and heated debate, and only after a national consultation of the membership through a series of regional meetings had taken place. In its policy statement, the BDA noted that "the drive to 'normalize' Deaf people, by increasing the quantity of sound which can be sensed, carries with it the danger of alienating the Deaf person from their own self-identity, and from their own natural community and its living language, without allowing full integration and access to hearing society."[15] It went on to argue that, while it supported both " the rights of adults . . . to choose cochlear implant surgery" as well as "the rights of parents (having received *all* relevant information beforehand) to act in the best interests of their child," the BDA was "unable to recommend cochlear implants for children." This mild statement was far from the rhetoric of Harlan Lane or Paddy Ladd, or the position that the National Association of the Deaf in the United States had adopted at the time. The BDA stressed the importance of parents being provided with adequate information regarding the Deaf community and the value of sign language for a deaf child. The cautious language and difficulty in reaching an agreed position are not hard to understand. Their diverse memberships and variety of concerns constrained the rhetoric used by representative bodies like the BDA. Late-deafened adults, a significant part of their memberships, tended to resist radical statements, and many deaf people were still fearful of public attention. It takes a degree of self-confidence to engage in public controversy, and many deaf organizations lacked the necessary confidence in the information they had available or in their ability to make their case convincingly in public.

In the mid-1990s, as the three years of earmarked funding from the central government were coming to an end, the MRC's Institute of Hearing Research was commissioned to evaluate the national cochlear implant program. Its report documents the growth of implantation in the early 1990s.[16] Seventy-four percent of adult patients (above the age of eighteen years) had received a Nucleus device as had 96 percent of child patients. Twelve percent of adult patients and 2 percent of children had received the Ineraid device that had been withdrawn from the market by the time of the report; and 10 percent of adults had received the UCH/RNID system. It was estimated that by the end of 1995, some six hundred adults would have received an implant, of whom three hundred would have been funded by the national program. "Implantation of children in the UK underwent radical changes of attitude and scale during the course of the evaluation. In 1990 only [three programs] accepted children . . . By the end of 1994, there were 15 centers providing implantation services to children. It had come to be considered appropriate to implant children as young as two years of age and to implant children with congenital hearing loss."[17] FDA approval of the Nucleus 22 system for use with children was said to have spurred these changes. Whereas at the end of 1990 only eighteen children (seventeen years old and younger) had been implanted, this figure was expected to reach the four hundred mark by the end of 1995. Considerable growth was predicted, in view of the backlog of demand, although "the future recurrent demand for CI services in the longer term will come predominantly from young children."[18] This 340-page report does not contain a single reference to the concerns expressed by the British Deaf community or even to the fact that the procedure had become controversial.

THE NETHERLANDS

In the Netherlands, as in Britain, initiatives to establish cochlear implant programs emerged at the grassroots level, with central funding becoming available only later. A crucial difference from the British case, however, was that the first implant operations were funded from existing hospital budgets, and so without appeals to local communities. Media coverage, which had been so essential to obtaining charitable donations in Britain, played a much smaller role in the Netherlands. Two programs began in the 1980s. One of these was at Utrecht University Hospital, where Egbert Huizing had been appointed professor of ENT in 1979. Huizing had been intrigued by a lecture by William House that he had heard some years previously, but it was only after his appointment in Utrecht that he felt himself in a position to put his plans into action. By 1982, Huizing had the material resources and the personnel he felt he needed to launch an implant program.

After three years of planning and information gathering, funding for three implantations per year were made available by the hospital administration. Because in 1984 there was much more experience worldwide with the single-channel 3M/House device than with any other, and because it was the only one to have received FDA approval, this was the device they chose to use.[19] The next step was to find potential candidates. With some reservations, it was decided to make the plans known through the press. From the people who contacted them, just a few were found to meet the selection criteria they had established. From these three were selected, and the first implant operations in the Netherlands took place in February 1985.

By this time a second program had started in the Netherlands: a collaboration between Nijmegen University's St. Radboud Hospital and the Institute for the Deaf, St. Michielsgestel, a well-known (then oral) school for deaf children. The beginnings of this program were quite different from that in Utrecht. The Nijmegen/St. Michielsgestel group chose to build up experience gradually. Their idea was to first develop a program of rehabilitation: teaching implantees to make sense of the unfamiliar sounds they heard with the aid of the device. Not only would the surgery be done abroad, they already had a list of people whom they considered potential candidates. However, the patients implanted abroad (in France and Germany) did not do well in rehabilitation. After a while, ENT professor Paul van den Broek and his colleagues decided that they would do the implantation surgery themselves. Whereas their Utrecht colleagues had chosen to work with the 3M/House implant, the Nijmegen surgeons opted for the single-channel 3M/Vienna device, on the grounds that this extracochlear device did not put any residual hearing at risk. When the hospital agreed to provide resources for the first six operations, implantation could move forward.

In the years following, cochlear implantation in the Netherlands was funded through a new national program, the Fund for Investigative Medicine. Projects approved for support were to be of an evaluative nature, with the patients tracked for three years so their results would help policy makers decide whether the technique had proven itself and should be routinely reimbursed. This was different from the situation in the United Kingdom. Although there too an evaluation was carried out (by the MRC Institute of Hearing Research), the funding provided did not depend on its results. Moreover, it covered both adults and children whereas that in the Netherlands was initially limited to adults.

From a policy perspective, cochlear implantation in the Netherlands was thus treated as experimental in the late 1980s and early 1990s. The purpose of the fund in supporting the two implant teams was to provide the assessments needed for political decision making. By keeping it in the realm of experiment, any political decision, with major resource implications, about reimbursement could be avoided. This did not mean that the implantation teams themselves regarded it as experimental. They had long since been convinced of its value. And they wanted to get it *on* the political agenda.

By 1991, five children had been implanted in the Netherlands. Dutch surgeons were well aware of the reactions that Chouard had encountered in the 1970s and of the objections of the NDCS in Britain. Opposition to child implantation was expected. As one surgeon explained, "My point of view is simply to avoid the opposition . . . Not everyone agrees with me. Some people say 'We have to convince the deaf world.' I am not going to engage in discussion with the Board of the Federation of Parents of Deaf Children. . . . Especially in the organized deaf world there is opposition . . . It is well known that it's this sort of people who go to congresses with banners saying *House keep your hands of my children*. Just as earlier with adults, we simply want it to become known in the Netherlands that we are almost ready to start with children. And then we think there'll be parents who'll come along and say 'we're interested.'"

In fact, there was no public expression of anxiety, let alone opposition to cochlear implantation in the Netherlands. Though leading members of the parents' organization (the Federatie van Organisaties van Ouders van Dove Kinderen, FODOK), were concerned, and in private some debate took place, advice to parents aimed at balance. Feeling itself inadequately informed and dependent on the medical and rehabilitation professionals for information, the FODOK avoided public expressions of dissent. Nor did the Deaf community in the Netherlands pay much attention to the issue, despite occasional bursts of publicity.[20] Indeed as of 1991, the Dutch Deaf community had paid no attention whatever to cochlear implantation.

This began to change in 1993, the year in which a new program of pediatric implantation began. Again funded by the Fund for Investigative Medicine, its beginnings were accompanied by a small flurry of publicity. For example, in *Trefpunt*, a periodical produced by the Dutch Ministry of Health, an article appeared under the headline "Electrical Inner Ear Beats Sign Language."[21] But the response of the Deaf community in the Netherlands remained lethargic, and as late as 1995 one of its leading members reflected sorrowfully, "While in the neighboring countries protest marches are held against the mental child abuse of cochlear implantation, nothing is heard from the Dutch Deaf community. While in other countries sign language is recognized and bilingual education already exists, it is hearing people who take the lead in the Netherlands . . . We, Deaf people in the Netherlands, do not adequately represent the interests of Deaf people and we do not adequately maintain our culture."[22]

Unlike the report of the British MRC researchers, the final report of the Dutch pediatric implantation program, published in 1996, did refer to the views of the Deaf community and the controversial nature of the procedure. It stressed that the Dutch centers "regard continuation of the dialogue with the deaf community as of great importance."[23] However, this apparent commitment to dialogue did not mean that the views of the Deaf community were allowed to influence the report's recommendations regarding the selection or rehabilitation of deaf children.

SWEDEN

As in Britain and the Netherlands, moves to begin cochlear implantation in Sweden also date from the early 1980s. But what happened in Sweden was not the result of independent and uncoordinated local initiatives. Instead of acting locally, Swedish ear surgeons and audiologists turned to the Swedish Institute for the Handicapped. This organization agreed to initiate a study of progress in the cochlear implant field. Then, in 1983, the Swedish Medical Research Council sponsored a conference on cochlear implants to which a number of foreign experts were invited. At the end of the conference, it was agreed to set up a trial project involving the department of audiology at Stockholm's South Hospital and the department of speech communication and music acoustics at the Royal Institute of Technology. Ten postlingually deafened adults would be implanted over a two-year period (1984–1986) and the results carefully evaluated. One year was set aside for this evaluation, and no further implantations would be done until it had been completed. As in the Netherlands, the surgeon involved, Goran Bredberg, was able to find resources from a reallocation within the hospital's existing budget.

The Swedes decided to use an extracochlear device because of its less invasive character, its lower cost, and because "the original study group decided that better results were not necessarily obtained with intracochlear or even multichannel systems." While, unlike his British colleagues, Bredberg was not forced to look to the community for funding, finding patients was problematic. He tried personal approaches first: "We . . . contacted patients who had been in our very hard of hearing program . . . we also contacted young people (now about twenty years of age) who had been through the school for the hard of hearing . . . tried to pick up those who had progressed to deafness."[24]

But of the people who approached them, four out of five were unsuitable, and finding suitable implant candidates was more of a problem than it was for the British or Dutch implant groups at the same time. Crucial in Sweden, and quite different from the situation in the Netherlands or the United Kingdom, was opposition from the Deaf community. "They were negative, and they said

. . . you know, in Sweden they have a very strong organization of the hard of hearing. They said, they should learn sign, they should learn the mouth-hand system where you give . . . Cued speech,[25] that's right. And they wanted the organization to have as many members as possible. And they said, also, this is artificial to have this kind of thing, this electric thing implanted, it's not natural. They said change the other things. And they said it's risky. They had a lot of arguments."

Vitally important to the success of the trial, according to Bredberg, was the positive publicity that successfully implanted people could provide: "Some of the patients who had been operated on, they had gone to the local press. They had gone to the weekly magazines and so on . . . there were a number of interviews . . . Which then built up pressure from the patients themselves . . . if we didn't have that contact it would have taken a long time."

In March 1988, the evaluation was completed and the results were presented at a meeting of the Swedish Medical Association. Since results were satisfactory and since professional opposition seemed to have abated, it was decided to continue the procedure and, where appropriate, use the twenty-two-channel Nucleus device as most other centers were doing. By April 1989, a further six patients had been implanted. In 1990, a second program was established, in the southern Swedish city of Lund. All this was done without any central government funding. Government funding for health care had been decentralized to the Swedish regions, and there were no central funds to be made available. Bredberg felt that the British strategy would not have worked in Sweden. He and his colleagues "have had no such contacts in high political circles."

In Sweden, the Deaf community did not hear of cochlear implantation through sensational accounts in the daily press.[26] Discussion of the implant began in the organizations of Swedish deaf and hard of hearing people well before the country's first implant operation was even carried out. In February 1982, *Auris*, the journal of the hard of hearing association HfR (representing *deafened* people, distinguished from the deaf not by the extent of their auditory deficit but by the timing of its onset) carried an article on research going on abroad. It noted that a working group had been established to follow events abroad. In mid-1983, when the decision to implant ten Swedish people was taken, this fact was also announced. More information followed: on the possibilities and the (clearly set out) limitations of the implant. The organizations were preparing themselves well before any operation had been conducted in Sweden. HfR asked to be involved in evaluation of that first series of ten implantations before any further decision was taken about the future of the procedure.

When the first operations took place in Sweden, they received the same sort of coverage from the daily press as they had elsewhere. But though the headlines were comparable with those found elsewhere, the texts were more restrained. For example, in December 1986, the wide-circulation newspaper *Aftonbladet* carried an article on thirty-one-year-old Asa Bergman-Blom, who had been deafened for four years ("I have got half of my life back, and my husband"). Surgeon Goran Bredberg was also interviewed in the newspaper. He stressed that although the implant gives a perception of sound, only some implantees would be able to recognize speech with its aid. Hearing would never be the same as before the onset of deafness. By this time, the Deaf organization SDR was beginning to consider the potential risks and benefits of the cochlear implant. In 1985, their periodical *SDR-Kontakt* carried an article on the possible dangers implied by the device. "The greatest danger with these 'machines' is that you are cheated into believing that the deaf has become a hearing person. Oralism has once again got the wind in its sails, with false arguments." The argument was a familiar one. It was being used by Paddy Ladd in Britain at the same time. However, this was not the important feature of the introduction of cochlear implantation in Sweden, what contrasted it with Britain and the Netherlands. It is the fact that the deaf and hard of hearing organizations had begun seriously to consider the implications of the device before implantation had even begun. What is more, even in publicizing the results of his operation, Bredberg did not make—and was not quoted as making—extravagant claims on behalf of cochlear implantation. The sources of Swedish difference here go deeper. They reflect the position of the Deaf in the country at the time the technology was being introduced, a significant point that is discussed further below.

Had the Deaf organizations sought publicity for their views? "I don't remember if it was in the newspapers. There was I think in the deaf newspapers . . . strong arguments against it. But the tougher discussion has happened a little later, when we started operating children. There they have made an official stand," Bredberg explained. "They have said no child should be operated before the age of eighteen, when he himself can choose . . . They made an official statement and they tried to write to all politicians."

By late 1991, three Swedish children had been implanted, not in Stockholm but in Lund. Aghast, the associations of the deaf, hard of hearing, and of parents of deaf children tried to organize a discussion with the management of the Lund Hospital. When nothing much came of that, they contacted the ethical committee of the Malmöhus county council, under whose jurisdiction the hospital fell. There was a sense among the Swedish deaf organizations that the medical teams in Stockholm and Lund were unwilling to engage in dialogue and that their views were receiving insufficient attention. Nevertheless, Bredberg and his colleagues had to take the concerns of the deaf far more seriously than did their colleagues in other countries. For a Swedish surgeon to have decided to "ignore the opposition," as his Dutch colleague did, would have been impossible.

> We have a bit of a careful feeling in Sweden about this, especially as to children. We have very good schools for the deaf, so that they sign. And there is a very strong organization . . . The children are very well taken care of . . . and they have an identity as "signing deaf" and also the sign language. Sweden accepted that as an official minority language, so that they have the right to translators, the right to be respected, and so on. This has made the alternative, you know, to operate . . . If you start with post-lingual children, already that is a big discussion going on, and a big trauma to these organizations. We have followed the literature. Meningitis children have very good results if you operate them, therefore we wanted to get started with that. But they said "no." But I talked to them, talked to the parents. If we go on their line the *child* does not have a real choice. Because at the age of eighteen, if you are deaf at the age of two or three you will have no possibility . . . you will also have no need, because you'll have an identity as a Deaf. You're already separated from your parents, and so on. You'll be in Deaf society. So in order to have the children have a real opportunity to choose, we must let the parents get all the information, not only from the deaf organizations, but also from us.

Bredberg had recently carried out an implant operation on a four-year-old child who had become deaf at the age of three. The child attended, and would continue to attend, a school that used sign language. By this time, deaf education in Sweden was largely using sign language and had been doing so for a decade. "This is a bit of a difficulty for us," Bredberg explained, "because if you look in the literature they say you should have them in the auditory situation, otherwise there'll be no good result. But I'm not convinced of that . . . the problem of the cochlear implant is very much a pedagogical problem." Bredberg had no plans at that time for implanting prelingually deaf children, partly to avoid escalating conflict with the organizations of the deaf.

Arne Risberg, a highly respected professor of speech technology at the Swedish Royal Institute of Technology, explained why he thought implantation of children had begun in Sweden.[27] After all, the schools for deaf children had all turned to sign language by then. Risberg was convinced that this had been beneficial, with one caveat. He thought that the schools were going dangerously far in abandoning spoken Swedish. "Parents of course like to be able to speak with their children, and like that their children be able to speak. So now, when they see that sign language is taking over completely they—parents—start to push for a cochlear implant. They think, wrongly, that their child will then be a speaking child. So in my opinion there is a very dangerous situation which has come up now." Pressure came especially from parents who did not share so much of Swedish culture: "This is the problem for physicians. Parents . . . very often of immigrants . . . they seem to be more willing to implant their children. Here in Sweden we had had a discussion about deafness for the last twenty years. And sign language has been accepted. So it's nothing special to be deaf . . . Deafness is accepted. But not among the immigrants. So they are pushing for implantation. Also there is one immigrant who sent their daughter for implantation in France. They tried to stop it . . . But the parents paid and they couldn't do anything."

The beginnings of cochlear implantation in Sweden were distinctive in a number of respects. The initial trial, conducted in a single Stockholm hospital and with a subsequent moratorium while results were evaluated, presented a very different starting situation from the uncoordinated local programs through which cochlear implantation began in most European countries. Programs in other Swedish cities (Lund, then Linköping) began much later. There was much greater restraint: in beginning pediatric implantation, in the way in which implantation operations were reported in the mass media, and so on. And finally, despite the feeling in the deaf community that that their views were not sufficiently respected, Swedish surgeons could not and did not ignore their views in the ways that colleagues elsewhere were doing.

A EUROPEAN OVERVIEW

By the mid-1990s, there were industry-supported moves to try to create European guidelines on cochlear implantation. Meetings with the department on "integration of the disabled" of the European Commission were arranged. Recognizing that views were divided, the head of this department agreed to sponsor a working group to be established by what was then the European Community Regional Secretariat of the World Federation of the Deaf (now the European Union of the Deaf, or EUD). The working group was asked to look into the social and medical implications of cochlear implants for deaf babies and children and into the quality of the social and medical follow-up within the different European Union member states. Chaired by Johan Wesemann (then director general of the EUD), the working group included surgeons, educators, and representatives of both deaf organizations and organizations of parents of deaf children. The working group sent a questionnaire to all the centers in the EU engaged in implanting children as well as to national associations of the deaf (NADs) and of parents of deaf children. Responses showed how rapidly pediatric implantation had grown in Europe, but also how unevenly. The two implantations reported between 1983 and 1986 had risen to forty-six in 1987–1989, to more than three hundred by 1990–1992, and to more than six hundred in 1993–1994. Moreover, of the thousand or so pediatric implant operations that were reported in total, fully 30 percent had been performed at one center: in Hannover, Germany. Four centers (in Germany, Britain, France, and the Netherlands) accounted for more than 50 percent of all the children in Europe to have been implanted. How pediatric cochlear implantation was financed differed enormously between the ten European countries in which it took place. In Britain and Germany, all aspects of the intervention were funded from normal health insurance funds. In France, a wide variety of funds were used, so not uncommonly a hospital used its own funds to purchase the device, while health insurance covered surgery and rehabilitation. In Belgium, the government had approved four devices,[28] but a surgeon had to apply to the central health insurance body for each patient to be implanted. About half of the responding centers required attendance of the child at an oral school, while for the other half education at a bilingual school or one using total communication was "acceptable." A few centers (but far from all) either routinely or on request would put a family in touch with the family of an implanted child, virtually none took steps to introduce families to signing deaf people.

Almost nowhere were there regular contacts between implant teams and associations of the deaf. The report of the working group summarized the responses of the deaf associations on this point as follows: "With the exception of Sweden, almost all respondents report that contacts with implantation teams are sporadic, informal, occasional, or even non-existent. In the case of Germany, the NAD reports that implant teams have explicitly rejected dialogue. Yet all believe that dialogue would be—or could be—helpful in potentially reaching consensus over eligibility for a cochlear implant. The Swedish NAD—reflecting on actual experience of consultation—stresses the value of this."[29] In the opinion of the working group, two things almost everywhere stood in the way of the dialogue that it felt of vital importance. One was a sense on the part of leaders of the Deaf community that clinicians "have a flawed and inadequate understanding of what deafness is, and are ill-equipped to advise on the best interests of a deaf child." The second was clinicians' unwillingness to debate with laypeople what they viewed as a matter purely of clinical practice. "If consensus is not achieved," the report states, "if compromises are not made, cochlear implantation will remain a source of conflict and potential distress."

Channels for Globalization of the Technology

The beginnings of routine implantation in these European countries, in the early to mid-1980s, were typically initiated by individual ENT surgeons, inspired by reports of what House had done by visiting Los Angeles and seeing him operate or by hearing him lecture at an international conference. Each of them then set about mobilizing resources in ways that depended on local and national circumstances. Why did they start? Questioned privately, a number of surgeons referred to personal motives quite different from the justifications used in public. A few explained that personal career considerations had played a part. Cochlear implantation was becoming a high-status procedure in the world of otolaryngology: a "high tech" intervention for the specialty. In the world of academic medicine, where scientific publications count towards promotion, to be active in such a "promising" area was a sensible career move. Other surgeons referred to the interests of their hospital in an increasingly competitive environment or to pressure exerted on them from the community. "Why aren't you doing anything . . . ?" Gradually, throughout the 1990s, the initiative passed away from these local physician-entrepreneurs. Channels of globalization were evolving.

Advocates of Deaf culture argue that cochlear implantation has to be seen as a "re-medicalization" of deafness. Insofar as this is so—and begging the question of how far deafness ever became de-medicalized in much of the world[30]—it is clear that medical professionals are no longer uniquely responsible for the phenomenon. Technologies that can be seen as vehicles of medicalization (of erectile dysfunction, for example, or infertility, or what is now known as social anxiety disorder), are today propagated principally by market forces rather than by the medical profession. "It is not new knowledge or technology that engenders medicalization but how they are used. Corporate and medical promotion of products, treatments, and drugs underlies the emergence of new medical markets. With our corporatized medical-industrial complex, the creation or expansion of medical markets becomes an important conduit to medicalization. Consumer demand is not simply unfettered desire for medical solutions, but it is shaped by the availability and accessibility of medical interventions," write sociologists Peter Conrad and Valerie Leiter.[31]

So far as cochlear implantation is concerned, processes of diffusion of the technology have evolved in two ways. In Sweden, at least according to one well-informed observer, parents of deaf children had helped initiate pediatric implantation. In the Czech republic, parents even more clearly played an important entrepreneurial role. In 1992, a young boy whose mother taught economics at Prague's Charles University became deaf after serious illness.[32] Seeking advice from all sides, the mother was ultimately advised to have the child implanted at the University Hospital in Hannover (Germany), the largest implantation center in Europe. When the ministry of health refused to pay for the operation, the child's mother launched a public appeal that succeeded in raising the funds. Within a year, moves to establish an implantation center in the Czech Republic were underway, and by 1994 two centers had been established in Prague. Although one was supposed to treat children and the other adults, in practice pediatric implantation soon dominated. A doctor from one of these centers secured a monopoly right to import the Nucleus device. Ernst Lehnhardt, who had established the Hannover implantation program, trained Czech surgeons to carry out the operation and supervised the first operations personally.

Having relatively recently emerged from a half century of communist government, the Czech Republic had little experience of voluntary organization. Nevertheless, by the middle of 1994, a Cochlear Implant Users Association (SUKI) had been established that broke away from the organization of parents of deaf children. SUKI quickly became an influential participant in Czech discussions of cochlear implantation. Lehnhardt was a regular speaker at their meetings. He offered a very clear message. All deaf children should be implanted, as young as possible, and irrespective of what the Deaf community might say.

Within three years, some thirty children had been implanted. However, the system of cochlear implantation that had been set in place so quickly was under pressure. This was pressure of a very different sort from that with which Swedish surgeons had to contend. Like deaf organizations elsewhere, the Czech Association of the Deaf—and a very few parents of deaf children—were opposed to pediatric implantation. Their arguments were the same ones as elsewhere. But in the Czech Republic, these arguments received little serious attention in society at large. To the contrary, the notion of the implant as a true miracle, a means of making a deaf child a hearing child, was the message that monopolized the attention of the mass media. Pressure on the implant system thus did not come from the Deaf community. It came from the much larger group of parents, and from SUKI, arguing that resources needed to be expanded: that the ten children per year for whom funding was available from public sources were too few.

Consumer participation in the work of translating implant technology from one "local world" to another is one element in the transformation of the mechanisms of globalization of the technology. Networks of medical training and professional collegiality provide another.

Despite the costs of the procedure, by the mid-1990s cochlear implant programs were being established in many countries of the developing world. Western-trained ENT surgeons and audiologists returned home with not only their professional skills but no less importantly a belief in the value of the technology. In some cases, they were able to secure access to government funds, however limited. In other cases, rich elite families provided an adequate base of paying consumers.[33] By the mid-1990s, cochlear implant programs had been established in much of Latin America, as well as in Asia, the Middle East, and elsewhere.

Pakistan, with an annual per capita income of six hundred dollars, is one of the world's poorest countries. Audiology services for people with hearing deficits are extremely limited and hearing aids have to be purchased privately. Few can afford them. Nevertheless, in 2000 a cochlear implant program was established in the city of Lahore. Advice and support were provided by the implant team at the University of Manchester, where senior staff had been trained.[34] Reflecting on the importance of the support provided from Manchester, the Pakistani surgeons and their British colleagues concluded that it had been crucial, that such "'dovetailing' of the local professionals with members of an established programme are crucial to set up a cochlear implant programme of the highest standard and these help to earn the confidence of other local professionals."

A more recent example of the functioning of professional networks in poor countries is in Uganda. In February 2008, a team of American specialists and nurses, led by a surgeon from the New York University School of Medicine and assisted by the head of ENT at a Ugandan hospital, carried out what was said to be the first cochlear implant operation in Africa.[35] A report on a Ugandan website explains that "after the operation, the patient was taken to the ICT department at Makerere University, where the implanted device was programmed over the Internet by another US-based specialist." Virtually none of the country's deaf people is likely to have access to a device that costs $26,000 (with surgery and rehabilitation doubling the cost). Yet national pride speaks loudly from this report. Representing as it does advanced technology, progress, and modernity, the importance of the cochlear implant for a poor country like Uganda is above all symbolic.

Channels for the Globalization of Protest

The channels through which a medical technology moves from one local world to another are those of the market, sustained by the promotion of consumer demand and by networks of medical collegiality and professional training. The "global" is invoked in support of local initiative, so that the Pakistani implant surgeons could explain the beginnings of their program by noting that the technology's "efficacy, safety, and reliability are well recognized," and that "over 100,000 patients have received cochlear implants worldwide." Starting in the 1970s, advocates of a sociocultural view of deafness in the United States, Britain, France, and a few other countries had developed a very different view of implantation. As discussed in the previous chapter, this was grounded partly on a critique of the evidence for the benefits of implantation and partly on a different reading of the experience of deafness. One aspect of this rereading was a new sense of the history of deaf people, one now seen as marked by a century of suppression of sign language and Deaf culture. The other was a renewed sense of community based on a shared experience of exclusion and stigmatization and on the use of sign language. In this context, cochlear implantation, and particularly the claim that it made deaf children hearing, was reframed in ethical terms, as a threat to the rights of deaf people and Deaf communities. The channels though which this critical discourse was propagated were very different to those available to proponents of implantation.

The channels through which critique was principally spread were those linking national associations of the deaf in a global umbrella body, Internet-based mailing lists used by individuals (including notably Deaf-L), and small scholarly communities focused on sign language and deaf studies.[36] Since critique was in part formulated in terms of bioethics, this field might have provided an additional avenue for the propagation of arguments critical of implantation. In the early 1990s, bioethics was taking root in many parts of the world, with an International Association of Bioethics established in 1992.[37] However, the bioethics that was being established around the globe was dominated by the principlism that also dominated the field in the United States. With its emphasis on individual rights and autonomy and its neglect of connectedness and community, principlist bioethics provided little support for the concerns of the deaf.

In brief, the channels through which a discourse critical of (pediatric) implantation spread lacked the resources and the capacities of those through which the implant and implantation practices spread. Despite this, the World Federation of the Deaf had adopted a critical stance in the early 1990s and the arguments were widely known among Deaf community leaders and activists. The major problem these activists faced in trying to contest the spread of implantation was not access to the arguments. It was more importantly access to the means of deploying those arguments in a specific national context.

Putting Technology and Critique to Work in the Local Context

The surgeons who established the first routine implantation programs in Britain, the Netherlands, and Sweden in the early 1980s were pioneers in their national contexts; they were social innovators. Each was faced with the challenge of assembling the financial, material, human, and symbolic resources needed to establish an implant program. How precisely this was to be accomplished depended very much on the funding and organization of health care in the specific setting. In one country, it involved negotiating a reallocation of the hospital budget, in another lobbying the national government or appealing to the local community for funds. In many cases, the mass media had to be mobilized in order to ensure a sufficient supply of potential implant candidates. Nevertheless, in meeting the challenge, these senior medical specialists were able to make use of well-established contacts with policy makers and with the media that their hospitals and professional organizations had developed over many years.

A decade or two later, with the establishment of implant programs in poor countries, matters were rather different. In Pakistan, for example, no government funds were available, and the program was established in the private sector and thus aimed exclusively at a wealthy elite. The social, economic, and cultural context, so different from the West, had a major influence on the way in which the program operated. Proof of adequate financial resources and easy access to one of the two or three cities in which rehabilitation could be offered were essential criteria in being admitted to the program. There were no standardized speech and language development tests in the country's two principal spoken languages (Punjabi and Urdu), so an ad hoc means of following patients' progress had to be devised.

Manufacturers also play an important role in supporting the establishment of implant programs in the poorer regions of the world. For example, in June 2007 Advanced Bionics entered into a partnership with the Hearing Solution Group ("Southeast Asia's largest hearing health care provider," serving Singapore, Malaysia, Indonesia, and Brunei).[38] In 2006, a Taiwanese philanthropist donated $270 million for the purchase of 15,000 Nucleus implants to be supplied to children in mainland China over a six-year period. The company and its Chinese distributor have together established a Cochlear Training and Education Centre in Beijing that "will train up to 500 habilitation therapists over the course of the donation period."[39] An "Awareness Network," modeled on practice elsewhere, has been established in the Asia Pacific region. These "volunteer advocates" "raise awareness about hearing loss and provide a solid support network for those with or considering a cochlear implant, including their families."

In contrast with the established channels for influence and the established health care institutions within which implantation practices could be developed, Deaf advocates had access to the channels of influence that had (or have) been created by and for the patient movement in some parts of the world. Patient organizations are said to be a new force in contemporary politics. As discussed in the first chapter of this book, their objectives are diverse. Some of them, in some parts of the world, have been successful in achieving influence over the allocation of resources for health care, or over priority-setting in medical research. Others have concerns that go beyond the spheres of medical or health care. Some deny the relevance of a medical framing of the interests of their members. The activism of people with disabilities, for example, seeks to reframe the needs of people with disabilities from one of medical rehabilitation to one of rights: the right to participate fully in social and economic life. These biosocial groupings, or "health-related social movement organizations," are just some of the actors in a transformed political landscape that has emerged over the past three or four decades. Others, including the women's movement, the environmental movement, gay liberation, and Black Power, were an inspiration to Deaf community leaders and advocates. Like the Deaf community, they too were as concerned with language and symbols as they were with traditional questions of economic justice. Like the disability rights movement, they too were struggling to "reframe" policies and attitudes. Despite the fact that the arguments available to these movements were the same everywhere, they were more successful in some countries than others. This fact has led some sociologists to conclude that relative success and failure has to be understood not in terms of the arguments alone but in terms of the ways in which political systems work and the different opportunities available to social movements in making their case and in influencing public opinion.[40]

Trying to achieve a comparable reframing of deafness, replacing a medical understanding with a cultural one, the Deaf community has been engaged in something similar. The cochlear implant had to be transformed from a symbol of medical progress into a symbol of cultural oppression and medical dominance. What social movement theory suggests is that their ability to reframe depends on much more than their arguments alone. It depends on the degree of their social organization, their access to the media and to influential people in society, and on the way in which the political process in one country or another works. In some countries, including the Czech Republic, the Netherlands, and the United Kingdom, deploying the arguments formulated by advocates did not happen easily. National organizations of the Deaf were hesitant (as in the Netherlands), had little or no access to the mass media (as in the Czech Republic), or had great difficulty in reconciling divergent viewpoints (as in the United Kingdom). The EUD survey suggested that in many countries, implant teams were unwilling to engage in dialogue with organizations of the Deaf.

In some places, clearer and more radical voices emerged, seeking not dialogue but confrontation. One of these was the French pressure group Sourds en Colère, founded in 1993.[41] Partly inspired by earlier gay liberation action, Sourds en Colère made opposition to cochlear implantation a principal objective: "The cochlear implant is experienced, within the deaf community, as yet another attempt at sociocultural genocide, of the same order as the banning of sign language at the Congress of Milan in 1880, bringing in its wake disastrous consequences for Deaf culture. But this time, the Deaf are not lowering their arms. Thus, *Sourds en Colère* are organising their first national demonstration, on 16 October, at Lyon, against the cochlear implant."[42]

The passion of Sourds en Colère, of which a prize-winning and nationally famous deaf actress (Emmanuelle Laborit) was a leading member, succeeded in winning attention for the deaf point of view.[43] In May 1994, a group of twenty (including representatives of Sourds en Colère, the sociologist Mottez, physicians, linguists, educators, psychologists, and parents of deaf children) presented a document to the French national committee on medical ethics (or CCNE).[44] They argued that given the uncertainties regarding the linguistic, psychological, and social implications of implanting deaf children, the technique should be regarded as experimental. They requested the committee to issue a ruling to that effect. Under French law, this would subject its use to rigorous control and oversight. A press conference was held to announce the document, and this resulted in a long article in *Le Monde* in which, almost for the first time in France, the concerns of the Deaf community received detailed and sympathetic coverage from a leading newspaper in an article titled "Language Quarrel among the Deaf. Supporters of sign language are concerned at the implantation of auditive prostheses in hard of hearing children."[45] In December 1994, the CCNE issued its report.[46] Though it rejected the claim that the cochlear implant should be regarded as experimental, its report greatly pleased the French Deaf community. The committee wrote that doubts regarding the precise benefits of the device were unlikely to be resolved in the near future. To avoid the possibility of compromising children's psychological and social development, they should all be offered sign language from an early age, whether they might subsequently become candidates for implantation or not. Thanks to the emergence of a radical advocacy group and to the existence of the national ethics committee, the Deaf community in France was better able to bring opposition to pediatric implantation to bear in the national context than were Deaf communities elsewhere.

In Sweden, too, the concerns of the Deaf community were taken seriously. But how that came about was very different from what happened in France. How was it that while the Dutch implant surgeon was able "simply to ignore the opposition," his Swedish counterpart felt obliged to take careful account of what the Deaf community in his country had to say?

The end of the 1960s had seen growing demands from Swedish deaf people—a national demonstration took place in late 1969—for greater access to social services, cultural activities, and education. Prompted by the National Association of the Deaf and the Government Commission on Handicapped Persons, in 1968 the Swedish parliament had passed a law accepting the right of deaf people to free interpreting services. Interpreter training courses were established, sign language research began at the University of Stockholm, a deaf theater group began, and signed programs were introduced to television. In 1976, the report of a government commission, "Culture for Everybody," argued that cultural policy should take greater account of the needs of hitherto neglected groups. The idea of using sign language in education was debated. Four years later Stockholm University established the first courses given in Swedish Sign Language. One year later came the most important development of all: the government's recognition of Swedish Sign Language as the language of the country's deaf minority. This presaged a rapid shift towards bilingual education in the country's schools for deaf children. The deaf organizations were accepted as representatives of this minority, with a right to be heard on matters affecting its well-being.

Countries differ greatly in their acknowledgment of the rights of cultural minorities. The French political scientist Dominique Schnapper has compared the ways different countries have tended to treat the cultural claims of immigrant groups.[47] Americans whose roots are in Greece or China are accepted as culturally distinctive, while enjoying full participation in American society as a cultural group. France does not recognize "groups" like that. You can only participate as an individual, and provided you learn French. As one might expect, therefore, in some countries organizations representing cultural groups are allowed a political role, while in others they are not. Schnapper suggests that Sweden is the extreme example of a country willing to acknowledge the rights of culturally distinctive groups. France is the opposite extreme: the country least willing to acknowledge any such rights. It seems likely that the rights granted to the deaf in Sweden—the right to education in sign language, to participation, to representation in political decision making—were a consequence of this more general feature of Swedish political culture.

The importance of sign language hardly needed stressing in the Swedish context. Educators were largely convinced that the use of sign language in school and preschool was in the best interests of deaf children. The Swedish deaf did not take to the streets or stage the symbolic destruction of an implant, as happened in France. They did not need to. They set out their objections to the implantation of children in carefully worded appeals to collective and personal rights, making use of the official channels of influence to which they had acquired access.

Globalization of a Contested Technology

By the beginning of the 1990s, the arguments against implantation of deaf children that Harlan Lane and others were continuing to develop were known to Deaf community leaders in all western European countries. But their ability to respond to this call to arms and the political opportunities available to them differed from one country to the other. Without recourse to the street theater of Sourds en Colère, the Swedish deaf had clearly achieved a great deal, from their point of view. Not only did the implantation of children start relatively late, but Swedish surgeons could not insist, as colleagues abroad often did, that an implanted child had to attend a school that used only spoken language.

There are major differences in the ways in which the implant technology on the one hand and the Deaf discourse opposing it on the other have been translated from one "local world" to another. The first has to do with the channels through which people, things, images, and ideas flow. The technology and its associated discourse and practices are propagated through market mechanisms, with the resources of major manufacturing corporations—resources that include the promotion of consumer advocacy, physician training, customer support, data collection, and so on—at their disposal. The discourse of the Deaf, by contrast, is propagated principally through academic journals, through Internet discussion groups, and through global gatherings of the Deaf community.[48] Despite the hopes vested in bioethics, even the institutions and international networks of this field could not be relied upon. Resources available here are vastly smaller.

The second difference has to do with processes of "localization": putting these flows of ideas, technologies, and practices to work in the local context. Market mechanisms are enhanced by professional collegiality and technical assistance as surgeons from developing countries travel to the West for training and return home imbued with not only the skills but the aspirations, values, and support of their erstwhile teachers. Western-trained surgeons, offering miracle cures based on advanced western technology, find receptive consumers among rich parents of deaf children in the developing world. By contrast, a well-organized Deaf community such as in Sweden, able to put the arguments developed by Harlan Lane, Paddy Ladd, and others to use in the local context, does not exist in much of the world.

A Pakistani ENT surgeon involved in establishing the implant program in that country wrote, "In a society like Pakistan, hearing disability is a stigma. There are cultural issues and issues of parental sensitivities. There is no well-defined deaf culture and deaf individuals not only face the dilemma of deafness but of cultural identity as well."[49] Fan-Gang Zeng, a professor of biomedical engineering and research director of otolaryngology at the University of California at Irvine, wrote of a visit he paid to China in 1993. During his visit, he asked many people, including deaf people, about the lack of a deaf culture in China. There is no deaf culture, according to his informants, because 80 percent of deaf children are unable to attend school because most deaf people are impoverished and because "signs used by deaf Chinese are not as fully developed as American Sign Language and signing is not uniformly recognized as a language."[50]

The accuracy of these observations regarding the lack of deaf cultures in the developing world cannot be taken for granted. Surgeons and biomedical engineers are not the best guides to understanding culture, and there has been little research on deaf communities or the emergence of a deaf identity in the non-western world. Yet there is undoubtedly some truth in what they say. Such historical and ethnographic studies as there are show great complexity and considerable difference between countries.[51] Thus the Japanese Deaf community seems to be marked by generational conflicts over the nature of deaf identity;[52] the Deaf communities of Thailand and Vietnam by major regional and local variations both in the sign languages used and in the adoption of a deaf identity;[53] while in Nicaragua the formation of a Deaf community (and a national sign language) has been shown to be quite recent, a result of the establishment of residential schools for deaf children by the Sandinista government.[54] These findings provide further support for the principal claim of this chapter. There are profound differences in the channels through which the technology and the discourse protesting its use flow, and there are profound differences in the structural possibilities for putting each to work in most local contexts. It is these differences that lie behind the globalization of implant technology.

Some thoughtful participants in the cochlear implant field are aware of these differences. In late 2001, a conference on pediatric cochlear implantation took place in Norway, organized by Oslo University's department of special education. Most of the hundred and twenty participants came from Scandinavia, and many were teachers of the deaf: experts in educational methods and in psychology. The Deaf community was well represented but there were only a handful of doctors. Most of the talks at this conference were different from those to be heard at medical conferences on cochlear implantation. There were few references to "miracle cures" or to futures without deafness. A Danish educator with many years' experience told of something he had recently observed. Two boys, both with implants, were communicating between themselves during and after school. During class, they used spoken Danish. In the break, the noise of the playground was such that they switched to sign. At the end of the day, walking to the tram, they once more used spoken language. But the clattering of the tram was too disturbing, and here they switched back to sign language once again. A Swedish colleague of his repeated a conversation he had had with an implanted child: "Are you deaf, or hard-of-hearing, or hearing?" The child responded, "I'm deaf, but I can hear with my c.i."

I too participated in this conference, and at it I made friends with two delegates from Iceland: Katrin and Anna. Both had backgrounds in the education of deaf children. Sitting together in the bar, one of them mused aloud, "There's such a lot that we here agree about . . . but we don't really get it across to people. It isn't our views or our experiences that really influence public opinion about cochlear implants." Katrin and Anna had seen that the channels through which information, technologies, and experiences are communicated are far from equal.

Are differences in flows and in localizations perceptible from within a specific local context? But also *how* are they perceptible within a specific local context? The anecdotes related at the conference introduce the question addressed in the following chapter. Using a different mode of inquiry, the focus is narrowed down to one country, the Netherlands. The chapter explores in detail how implantation began and the opportunities for putting the conflicting discourses to use as it evolved.

Chapter 5

Implantation Politics in the Netherlands

WHEN THE FIRST ATTEMPTS at developing a cochlear prosthesis, in the 1970s, became known in the Netherlands, responses were mixed. Some ENT surgeons, including Egbert Huizing, were intrigued and wanted to try it out. But others were more doubtful. The Amsterdam professor L.B.W. Jongkees, a leading figure in Dutch otology, wrote a highly skeptical piece in the country's major medical journal:

> Dr. Chouard from Paris has just favored the world with an indication of the remarkable success of French medical science by sending out an announcement that he and his colleagues, by computer-assisted signal amplification directly to the auditory nerve, are now in a position to let the deaf hear. The modest scholar adds: so far as hearing speech is concerned, matters are now resolved, though not yet for music. Poor deaf, poor family practitioners, poor ear doctors. It is a soap bubble . . . Without careful experiments—for which animals are far more appropriate than people—it is highly unlikely that through playing with electrodes, sticking them in the auditory nerve without much understanding, there is much more chance of providing a deaf person with useful hearing than there is of a rhesus monkey, provided with a typewriter, of producing the book of Genesis, even in Swahili.[1]

By the beginning of the 1980s, opposition was fading. In a 1982 article in the same journal, Jongkees admitted that, whatever the lack of fundamental understanding, the device did seem to help some at least among the deaf.[2] FDA approval of the 3M/House implant for use in deaf adults, in late 1984, reassured Dutch surgeons. If anything went wrong, they would be able to justify themselves by referring to the FDA approval.

It was at this time, the mid-1980s, that the two implant programs that constituted Dutch efforts in the field until the early 2000s began: one based at Utrecht University Hospital, the other a collaboration involving the university hospital of the Catholic University Nijmegen and a nearby school for deaf children called the Institute for the Deaf, St. Michielsgestel, or IvD.

Though these two programs developed along parallel lines, and were at times obliged to collaborate with one another, their origins were different. The Utrecht program, like many of those in other countries, had its origins in a hospital department of ENT. The Nijmegen/IvD program evolved out of the resolutely oral philosophy of the school. At the IvD, years after Stokoe's work on sign languages had attracted international recognition, the status and significance of these languages were ignored or rejected. The IvD, a Catholic foundation dating back to 1840, was at that time well-known for its exclusively oral approach and its firm opposition to the use of sign language. Toine van Uden, a priest and former director of the school, continued for many years to deny that sign languages were languages at all. Van Uden's educational methods, developed in the 1950s, were continually being refined with the aid of new technologies.

The Utrecht program began in 1984 after some years' preparation. The Utrecht group opted for the single channel 3M/House device because at the time it was the only device that had received FDA approval and there was much more experience with it, worldwide, than with any other. Huizing was able to convince the university hospital to establish a cochlear implant program, and funds for three implantations per year were found from within the hospital's budget. When all had been prepared, the next step was to find potential candidates: "I really wouldn't know how it can be made known generally, and in the deaf population, other than via the channels of the deaf magazines, the general press . . . I really wouldn't know. And we really needed that, otherwise we couldn't have started."[3]

Though public appeals for funds were not needed, the mass media still had an important role to play in informing potential implant candidates. It was recognized that publicity could be dangerous—perhaps deaf people in the community would expect too much—but there seemed to be no alternative:

> In our way of thinking that isn't something one readily does . . . we don't like it. We're always a bit scared of the press . . . many doctors are somewhat scared of the press. Because the press is always rather looking for the sensational, and tends to suggest more to the patients than the doctors can actually provide . . . A dilemma, this. On the one side you have to inform the public, on the other side you create unrealistic expectations . . . In this case we let ourselves

> be interviewed by the press . . . I really wouldn't know how you can inform the population, and the deaf population, other than via the deaf magazines, the general press . . . And we really needed that, otherwise we couldn't have started.[4]

In the two years following, eighty people contacted them about the possibilities of an implant. Most could be rejected immediately on the grounds that they did not meet the selection criteria. Careful selection criteria were important if the results were to be successful: only people above the age of twenty who had acquired a mastery of spoken and written Dutch before becoming deaf would be eligible. They would have to be in good health and have no other handicaps. A few people meeting the criteria were found, and in February 1985 the first implant operations took place in Utrecht. Three women had been selected: the oldest aged sixty-two years old and deafened thirty-eight years previously, the youngest twenty-nine years old and deaf for four years.

By this time, the second program in the Netherlands had started. The IvD, committed as it was to oral education, was interested in pursuing any new technology that might lead to better oral rehabilitation. It was this interest that in the early 1980s led the IvD to send a few of its former pupils abroad for implantation and to work at developing a rehabilitation program that they would follow when they returned. They had drawn up a list of people whom they thought likely candidates. In 1983 one man, whom they thought was having difficulties in coping with his deafness, was sent to Paris to have an implant fitted by Chouard. The result was a disappointment, and the implant had to be removed. Then one or two were sent to Düren, in Germany, to be implanted with a different system being developed there. The idea in each case was to build up experience with the work of rehabilitation: teaching the patients to make sense of the auditory experience the implant provided. But the second group of implantees was little happier with their implants than that first man had been. At this point, collaboration with the ENT department of the nearby Nijmegen University Hospital started. After a while the Nijmegen surgeon, Professor Paul van den Broek, decided that he and his colleagues would do the implantation themselves. They started developing selection criteria and language tests in the Dutch language and thinking about which device to implant. Differently from their colleagues in Utrecht, the Nijmegen group opted for the single-channel extracochlear 3M/Vienna device, on the grounds that the surgery involved was less invasive. When the hospital agreed to fund the first six implants, all was set.

In 1986, the Dutch Association of Audiology published a report setting out the situation in the country.[5] The reasons for doing so, the forward explains, have to do with the expectations among deaf and hard of hearing people to which sensational press reports had given rise. The hope was to create a more realistic climate of informed opinion. "As of today," the report states, "with the exception of a few 'star patients,' there is no electric inner ear prosthesis that restores the possibility of auditive speech perception to the wearer. It does provide valuable support for lip reading." Most of the sixty-three-page report dealt with experiences in Utrecht. Huizing described the origins of the Utrecht program and the variety of skills needed for a cochlear implantation program to function: an experienced ear surgeon to do the actual implantation, an audiologist (whose job it is to conduct the necessary measurements, regulate the implant, and help establish the revalidation program), a speech therapist to lead the revalidation program (and thus to work closely with the implantees), and a social worker to provide psychological and social support to the patient and his or her family. Huizing criticized the refusal of Dutch health insurers to pay for the device, though the operation itself and the rehabilitation process had been reimbursed. Using estimates made in other countries as a basis (estimates suggesting that approximately 10 percent of all deaf people would meet selection criteria), there were expected to be "a few hundred" suitable candidates in Holland. Other chapters described the various devices then available, patient selection, surgery, and the revalidation program. In a final chapter, Frans Coninx, head of the IvD's research and development department, described what had happened with a forty-four-year-old patient, deafened by an attack of meningitis at the age of seven, who had been sent to Germany for implantation. This patient himself was enthusiastic about his implant. He felt that his quality of life and his speech had improved. People around him, however, were mostly disappointed. They had expected much more.

In early 1988, the Nijmegen/IvD team was informed that the Health Insurance Council (Ziekenfondsraad) would provide support for a three-year project from its Fund for Investigative Medicine. This would allow them to carry out fourteen operations (in addition to the six that had already been done using the institutions' own resources). These fourteen operations took place between 1988 and the end of 1990. In early 1989, a leading newspaper published an interview with one of these implantees, a woman who had been deaf for more than forty years as a result of childhood meningitis, under the headline "Deaf Woman 'Astonished' to Hear Sounds Again after Implantation." This was the article that I clipped out, and that aroused such hope of a miracle cure for my son.

In August 1991, a report was submitted to the Health Insurance Council. It explained that after ten procedures making use of the Vienna single-channel extracochlear system, the Nijmegen/IvD team had switched to the Nucleus intracochlear device. This was because reports from abroad suggested that the more advanced coding strategy of the Nucleus led to better results. Although some centers abroad did not accept prelingually deaf people into their programs on the grounds that benefit was likely to be small, the Nijmegen/IvD team had decided to include these too, given their long experience in the rehabilitation of prelingually deaf people. More than three hundred people had approached them since the beginning of the program, mostly as a result of media publicity. Few had been referred by a medical practitioner or an audiologist. Of these three hundred plus, 43 percent had subsequently withdrawn themselves, while another 47 percent did not meet the selection criteria. Of the thirty or so remaining, twenty had been implanted, while the rest were still in the selection process. The people who had been implanted were not sign language users: nineteen of twenty implantees were wholly dependent on speech for communication.

What had the patients gained from their implants? In addition to the usual tests of speech perception and production, the patients implanted had also been asked to report on their subjective experiences. Though most noted an improvement in their ability to communicate verbally with others, about a third of them, mostly prelingually deaf, were disappointed in the results: they had expected more. "One can speak of an increased feeling of security on the street. This can be regarded as an important gain for the deaf. The treatment has made more contacts with hearing people possible, sometimes at the cost of contact with deaf people . . . One in four who was working indicated that their career chances had improved."[6] Despite the costs of the procedure (around $45,000), despite the fact that the prelingually deaf had largely been disappointed, and despite the fact that improved contact with hearing people was "sometimes at the cost of contact with deaf people," the implant team was convinced that "prelingually deaf patients can also profit substantially from a CI."

The implant team recognized that the success of the program depended crucially on suitable patients coming forwards and on the criteria used in selection. The fact that most people who had approached them were self-referred, rather than having been sent by a doctor or an audiologist, was a matter of concern since few of them came through the selection procedure. General practitioners and audiologists needed to be better informed regarding what implantation had to offer. Moreover, since "many deaf people rarely see an ENT doctor or visit an audiological center, information via deaf organizations and news media will be important." So far as selection criteria were concerned, they estimated that roughly 10 percent of deaf people in the Netherlands, or some 800 people in all, met the basic criterion of having no benefit from a conventional hearing aid. Roughly a third of these, between 250 and 300 people, were likely to meet the selection criteria they had established. Others were likely to have additional handicaps, or too much residual hearing, or be otherwise unsuitable. Foreign studies suggested that the benefit was greater in the case of deaf people who used only speech. This criterion halved the numbers, with the 150 or less people dividing more or less equally between pre- and postlingually deaf. The pool of candidates for whom good results were to be expected was thus extremely small. The major focus would have to be on children. "Children form the most important target group for CI. Implantation at a young age, in particular, seems vital since the negative effects of deafness on communicative possibilities will be less. The considerable care required, both before and after implantation, and the strong interaction with education, suggests the need for a special infrastructure."[7]

From October 1991 to October 1994, the Fund for Investigative Medicine supported a further trial of cochlear implantation in adults, in which the two teams were obliged to cooperate. Pooling of data would make for better analysis, since the numbers were so small, and this would only be feasible if the studies were comparable. But the real interest in each of the centers was in implanting children, and by 1992 they were developing their plans for doing so.

An Ethnographic Approach

It was at this time that my own work started. Interviewing some of the leading members of the two implant teams made clear that they were convinced of the value of the technology. They had anticipated opposition to implantation of deaf children in the Netherlands, as had been the case in other countries, but so far there had been none.

Doctors I interviewed wanted me to understand their view of the deaf and of the sources of deaf objections to cochlear implantation that had emerged abroad. Thus one senior otologist explained to me that the deaf world is actually two distinct worlds: the deaf born (growing up in a distinctive deaf culture), and the late deafened. The second is a smaller group, lacking the ties and the organization of the first group. People who become deaf in adult life are generally isolated and lonely. It is the former group who sets the scene, and who is "doubtful" regarding implantation: "*we know* that they have less to expect from cochlear implantation. So it's really a question, if you should offer it to them. Nevertheless, all the deaf born who have been implanted *assure me* that it contributes enormously to their social functioning. . . . it's not acceptable that the deaf world should be able to deny some of the deafened, who can profit enormously, the possibility of getting the implant . . . it's other sort of people."

Towards this interview's end, I introduced a note of mild dissent. I explained some of the reasons for which, I had read, Deaf communities oppose implantation, particularly in deaf children. Had they encountered significant opposition? "No, we've not noticed much on that score." Abroad, however—he'd made the point earlier—there is significant opposition. It is all to do with different ideas about deafness (the deaf rejecting the hearing world's view of them as handicapped) and with the endless discussion about "sign language versus spoken language . . . that plays a role here too, in everything." Still later, the tape recorder packed up, I mentioned that I was myself the father of a deaf child. I felt I needed to correct the impression I must have made. The reaction was one of surprise. I was immediately assigned a different but familiar status, that of interested parent. The surgeon I was interviewing shifted rapidly and easily to a different script. I was interrogated as to the exact nature of my child's deafness. Was I myself considering implantation?

Another senior Dutch doctor told me, "It will come, no one can stop it." Eventually, he was sure, all deaf children would have cochlear implants, "though I don't know if I'll be around to see it." Did everybody share this dream? I knew by now that they did not. For Deaf intellectuals like Paddy Ladd and Benjamin Bahan, for Harlan Lane, it was no dream but a nightmare. Their views on the world of the deaf are very different from what I had heard. The Dutch surgeons seemed to see deafness as a tragic affliction, and deaf individuals as sad and lonely figures. Admittedly, some found a kind of solace among others like themselves, but the task of medicine was to help those who aspired to enter the hearing world. The view was of a small group of deaf people who communicated principally through spoken language and of a sign-language-using Deaf community that sought to deny them their rights.

A discussion with Harlan Lane soon after these interviews forced me to reconsider what these Dutch clinicians had had to say. Lane told me of Stokoe's research and how it had begun to make clear that the gestural languages of the deaf were indeed full and natural languages. He explained how this sign language research was leading to greater self-confidence on the part of the Deaf community. From this perspective, it was not the repression of a small group of non-signing deafened people by a signing Deaf majority that was at issue but the historical repression of a cultural community by hearing society. From this perspective, the cochlear implant appeared as a weapon designed to reverse the gains that the Deaf community had made.

I began to see that I would have to change my approach. In order to understand what the introduction of cochlear implantation to the Netherlands had entailed, I would have to try to uncover whatever alternative understandings of the cochlear implant had emerged in the Netherlands, even though they had apparently not crystallized in public opposition. I would have to try to understand who was thinking about implantation, and how they interpreted what was going on.[8]

The first thing to be decided was "where to look"? Where might alternative understandings of the cochlear implant have emerged in the Netherlands? I had been told that there had been no opposition in this country. It might be quite different from what Lane had told me of the United States. Was opposition to the implant beginning to emerge in the Netherlands? Was it based on the kinds of arguments Lane had developed? Where might such opposition be crystallizing?

On the basis of experiences in other countries, organizations of the deaf and of parents of deaf children seemed the most likely places to look. Being myself a member of the organization of parents of deaf children, I felt reasonably confident of my access to this group. But access to the Deaf community was a more daunting proposition. As various hearing scholars have testified, it is not easy for a hearing researcher to study the Deaf community.[9]

I approached the national organization of parents of deaf children (FODOK) through an acquaintance who was a member of their executive board. The organization was willing to talk with me partly because I was (and am) part of their constituency. In December 1992, after delays that made me impatient, I was able to arrange a discussion with members of the FODOK executive board. This discussion enabled me to reconstruct the chronology and the basis of FODOK's involvement with the issue of cochlear implantation.

FODOK representatives had first heard about the technology in 1985, in the context of an international conference on the education of deaf children. They had been shocked by the technical nature of the discussion, with no attention paid to social and psychological issues. FODOK had come to the tentative conclusion that, given the experimental character of the procedure, implantation of deaf children should not be approved. However, in the Netherlands, at least, the issue had not yet arisen.

The topic emerged as a matter for discussion within FODOK at three distinctive moments, two of which had been reactive: triggered by the funding proposals of the implant teams. Ever since the issue had first arisen, FODOK had been conscious of a lack of independent knowledge and expertise on which to base its position. The result was that both in formulating its views and in the timing of their expression, they had found themselves reliant on the implant teams. (For example, an audiologist from the IvD, setting out on a study trip to the United States in 1991, had invited FODOK to provide a list of questions to which they wanted answers.) No independent policy had emerged. Certainly there had never been any question of the public expressions of concern that sister organizations in France and in Britain had articulated years earlier. Quite recently, FODOK had been asked to write a letter affirming their support of an application to the Fund for Investigative Medicine. In addition, the implant team was establishing a steering committee in connection with this project, and FODOK had been invited to nominate two representatives.

Consider now the discussion with FODOK that I had in December 1992. It was clear to everyone at the table that the discussion was taking place because I am both a sociologist studying medical technology and a member of the group, sharing the same concerns and interests. The discussion involved much more of me, of my life, than my status and professional responsibilities as a researcher. It constantly referred to my own children and the problems and dilemmas, shared with the others, of parenting deaf children. Interviewing clinicians had involved my background, too, but it had been subsidiary. I had referred to my children only at the end of an interview, fearful that otherwise I wouldn't get past the standard message given to parents. Here, by contrast, it was virtually a condition of the discussion taking place at all. Participating in the meeting were members of the FODOK executive board and one of the representatives on the Fund for Investigative Medicine project steering committee (which had met just once by this time). The latter, Jan, is himself the father of a deaf child who has received a cochlear implant.

As the first few children in the Netherlands were implanted, a source of knowledge became available to the FODOK. This was the experience that parents themselves accumulated through their interactions with their implanted children. Jan represented that experience. His claim to authority derived from the fact that he had daily contact with an implanted and growing child. It was he whom the FODOK had asked to represent them in the steering committee. Although I was not aware of it at the time, I was in effect challenging the authority of his knowledge. How should we think about the benefits of implantation for deaf children? Jan explained how, from his point of view, the implant had been of great value: "The fact that before I had to go upstairs to say 'come and eat,' whereas now I can just shout up . . . Life is just that bit easier now." Surely much more was at stake than that? I found myself opposing my own understanding to his: explaining why, in my view, we have to understand the implant's benefit, if any, in much more abstract terms than these. Thinking back to what Robert Ruben had said in Nottingham a few months previously, I said, "There has been no attention to language. The ear doctors, the audiologists . . . they know nothing of language." Significant here is the fact that this discussion was taking place under the auspices of FODOK. Perhaps I was trying to justify having taken an evening of everybody's time, trying to make clear that what I brought to the discussion was also of value. Convincing the people around the table that their evening had not been wasted, trying to give them something in return, became—in this context—an attempt at demonstrating the superiority of my own kind of knowledge to the knowledge based on personal experience to which they had access. In so doing, as I was soon to discover, I had crossed an important bridge. The reciprocity at which I had aimed had in fact succeeded in turning my own understanding into a potential resource.

The possible value of what I knew was already being weighed in the balance. At the end of the meeting, I was asked if I would be willing to work on a FODOK document, "guidelines" for policy. A little later, walking to the railway station, I chatted to a member of the FODOK executive board who began to reflect aloud. Had it been wise to have made Jan—deeply committed as he is—FODOK representative on the steering committee? Might I perhaps be willing to serve on the committee in his place? FODOK was considering shifting its reliance, in part at least, from the experiential knowledge represented by Jan to the scientific knowledge that I was implicitly claiming. In a committee that would surely be dominated by clinicians and professors, my statements on behalf of FODOK would have the advantage of that status, of scientific authority. After some hesitation, I accepted the invitation. Letters passed back and forth. In January 1993, the implant team was notified that I was to take Jan's place as FODOK representative. The reaction was not wholly positive. From their point of view, my opinions were an unknown quantity and thus potentially disruptive. If all this had taken place a few years previously, I might have justified what I was doing in terms of the familiar notion of helping FODOK effectively and independently articulate its views. Previously they had lacked a basis for grounding a distinctive parents' view. With the emergence of experience-based knowledge, that was no longer the case. So why accept? To be sure, there were potential advantages to me in terms of data collection. Continuing involvement with FODOK in this area could provide a valuable source of information. Had I come to mistrust the experience-based knowledge that Jan represented? Was this because I was concerned by the enormous emotional investment that lay behind it? Or was I making inappropriate use of the authority of scientific knowledge? This continued to nag.

My insights into the views of parents and their organization were not paralleled by any similar knowledge of how the Deaf community perceived implantation. Gaining access here would be a very different matter. My sign language skills were not adequate to interviewing deaf people on an issue like this. In the course of a social event organized by the local Deaf club for deaf children, chatting to some of the people present, I had an idea.

Couldn't I involve someone from the community as an assistant, something that ethnographers often do? Just as in my discussion with FODOK, I was present by virtue of a status other than that of sociologist. Access to the Deaf community, via an informant or assistant, followed from my status as a parent and from the responsibilities that the Deaf community felt towards its budding members.

I explained my research to a sympathetic and intelligent young man. Might he be willing to work with me? Johan was interested. I explained to him that I would like him to try to find answers to two sorts of questions: how interested has the Deaf Council (Dovenraad)[10] been in cochlear implantation, and since when? Do individual deaf people have any views on the possible implications of cochlear implantation for the Deaf community? The former question could be dealt with through a number of interviews with leading figures in the community, the latter through organizing one or two discussion meetings at local Deaf clubs. An article appeared in the national deaf magazine *Woord en Gebaar* with the headline "C.I. Study Begins at Last."[11] Johan's article informed readers about my project, about the need to learn of the views of deaf people, explained his role, and invited readers to contact him. A forthcoming national event, at which hundreds of deaf people would be present, would provide another opportunity for collecting information.

Some sixty people filled in a questionnaire that Johan handed out. This was not a representative sample by any means, but their comments gave us an idea of the range of opinion and knowledge among deaf people in the Netherlands. Some saw the implant principally as a threat, while others saw it as potentially valuable. Some spoke of what it might mean for them as individuals, others of what it might mean for the Deaf community. Some were worried by how the mass media praised the implant, others by the way in which they thought hearing parents dealt with their deaf children. There were many sources of anxiety, and there was a definite sense of injustice.

In answer to the question "Do you know what the disadvantages of CI might be?" someone wrote, "Yes, there certainly are disadvantages. For example, loss of identity, the inadequacies of the C.I., and the hearing parents who get too little information from the 'audist establishment,' and have far too high expectations."

In answer to the question "Have you ever met an implanted person? What did you think?" someone answered, "No. Only on TV. A hearing father made a remark that made me feel sick. He said, thanks to the 'favorable result' of his deaf son's implantation, 'Now I've got a normal child.'" Another respondent had met an implanted person: "A man and a woman, both of them hard of hearing people who were becoming deaf. Unfortunately they couldn't accept it, to continue to live without sound. I can understand that. If they had been deaf, I would have found it puzzling. I would have found them pathetic, in the sense that they wouldn't have accepted their deafness."

A deaf woman wrote, in answer to the question "Do you see CI as a threat?" "Yes, if CI is used on a large scale by deaf children under the age of 16. It means that parents are deciding for their deaf children, and I don't find that a good thing. Deaf adults can decide for themselves. Deaf children just have it done to them."

People understood the notion of threat in very different ways. "Yes. They drill a hole in your skull. Horrible," said one. A young woman with deaf parents didn't see it as a threat at all. "Let them try," she said. "There are new developments in the treatment of every sickness. To improve things. So why not us?" But this woman did not really know what the cochlear implant was and had only recently heard of it. The mother of young children who had gradually lost her own hearing answered the question of whether she saw the implant as a threat with "Yes and no. A threat for the Deaf world/culture (making deaf into hearing). Personally I find the idea of hearing sound again delightful."

The way the implant was presented by the media irritated many people: "Too positive. Lots of people get a very positive picture from it. They think you become hearing," and "Goodness, a great sensation. This new thing arouses great expectations in many people. It's just I think they had a right to know about the negative side, for example that you can't hear sound for 100%. To avoid disappointment." Another respondent wrote, "I find it scandalous that they never ask for the opinion of the Deaf. It has quite a lot to do with them too. It seems to me that the media (TV/press) are only interested in what CI means from the point of view of hearing people."

Although many of the people responding had views on the matter (or presumably they would not have taken the trouble to fill in the questionnaire), little or no formal discussion of cochlear implantation had taken place in the Deaf community prior to 1993. Unlike the parents' organization, the Deaf community had neither been invited to participate nor had felt any great need to formulate its views. In the course of 1993, and in parallel with our own study, this began to change.

Perhaps our work was making people more aware of the issue, focusing their attention on it. It sometimes seemed that way. Certainly, the discussions that Johan organized at Deaf clubs around the country were among the first formal discussions of what the implant might mean for the Dutch Deaf community. On October 12, 1993, one of these discussions took place at the Amsterdam Deaf Club (SWDA). This is what I wrote in my notebook some hours later:

> Because it was Anja's birthday we couldn't participate fully (as observers). Anyway, we took the video equipment there at 4:30, taking the children along with us. Jascha found it lovely to be there, as always, and just wanted to stay. (We agreed that we had to take him there from time to time). Then, at 7:50 (a bit later than intended) we went back to find things in full swing. It was by no means easy to follow. They'd set up the room as a circle of tables with Johan and Jenny in the front. SWDA was also making a video for their own archives. Discussion raged, and emotion clearly ran high. Johan found it difficult to control the meeting. Personal experiences, knowledge and information (or lack of them) . . . (Metty, who was there, told me that someone had thought you got a real hole in your head. How did you fill it up?) There were 20–30 people: a good turn out. But such a variety of personal experiences and emotional careers . . . Ideas about deaf identity were put forwards, and the need to give up adjusting to the hearing world.

This discussion evening, organized by Johan as a means of eliciting opinions (research data for us), had been significant for the Deaf community too. They too had recorded the discussion on video. It led to an article published in the Newsletter of the Amsterdam Deaf Club soon afterward.[12] What it had shown, concluded the author, was clear unanimity regarding implantation both of adults and of children. Deaf adults are able to decide for themselves whether the device is something for them or not. So far as children are concerned, "People found it a very bad thing if deaf born or early deafened children are deprived of their right to be Deaf, and if being Deaf is viewed so negatively by the hearing world." More broadly, the writer set out a Deaf perspective quite similar to what I had heard from Harlan Lane a year or so earlier. Choosing for a cochlear implant "means also abandoning your identity as a deaf person, and acquiring a new identity. Just as when a man, after hormone treatment and an operation, lets himself be turned into a woman. The difference is that in this (CI) case the majority stands cheering. The difference is also that as an ex-deaf person you don't become a hearing person."

For the first time too, the deaf perspective on cochlear implantation was noted in the mainstream press. In November, one newspaper carried an article that portrayed deaf people as finding it strange that a child should have to undergo a serious operation to correct something that hearing people—but not they—see as a handicap. Deaf people, after years of having been outsiders, have created their own world with their own culture. (American) clinicians seem not to understand this point of view, and draw an analogy that deaf people would not draw: "There aren't any blind people who'd protest if a new treatment was discovered that could improve sight."[13]

After the discussion evenings in Amsterdam and other Dutch cities had been concluded, Johan wrote a brief report giving his interpretation of what he had found. Here are some thoughts from his report.

> There are various views regarding CI. The general feeling about ENT specialists and about the hearing world is rather negative. What particularly struck me is that the Deaf accept that Deaf adults let themselves be implanted, but that they also say that those who do don't accept Deaf culture and their own identity. There is general agreement that the implant is good for people who become deaf. On the other hand there are strong feelings against the implantation of Deaf children . . . Hearing parents are always proud if their Deaf child acts "hearing." They never understand what Deaf is; they doubt if Dutch sign language offers enough for the future; their command of sign language is less than their child's; they don't realize how important Deaf culture is for good social and emotional development; they always want to do their best for their child . . . except learn sign language well . . . The Deaf know and accept their own "handicap." Deaf children can't explain to their parents what it feels like, that their parents chose an implant or a hearing aid, without letting them have any say in the matter at all. The Deaf fear that Deaf culture will slowly disappear. Some of them feel themselves the Last of the Mohicans. Others think that there will always be Deaf culture, because CI users will continue to come to the Deaf clubs, and because not all Deaf are suitable for implantation. . . . Worried about Deaf children, because hearing parents show their children a "false world." I know that from my own experience. For years I tried to live only in the hearing world and only later realized that your real world, life and culture is with the Deaf. That is true of hearing aid users too. Deaf is also your identity. Better totally deaf or totally hearing. Not something in between. . . . A sign language teacher gives NGT lessons to hearing parents of a deaf child. The hearing parents are really enthusiastic about sign language. The Deaf teacher feels proud. But subsequently the parents have their child implanted. The teacher feels frustrated. Other deaf people get goose pimples from it.

In this first period of my work, the introduction of cochlear implantation in the Netherlands was proceeding without overt controversy. There were no "sides," despite anxieties in the Deaf community and despite clinicians' expectations (based on the experience of foreign colleagues) that opposition would come. Because FODOK had been invited to support proposals for funding, the organization had been made very aware of its own lack of independent information and expertise. The organization felt a need to participate, and it wanted someone whom it could trust to articulate its views where necessary. I appeared to be such a person, and as explained earlier I decided to take on this responsibility. So far as the Deaf community was concerned, matters were different. It was only in 1993 that members of the community were beginning to formulate concerns along the lines that Harlan Lane had explicated, to some degree as a result of the work that Johan and I had been doing.

But not as a result of that alone, for 1993 also saw the beginning of the Nijmegen/IvD pediatric implantation program. Thanks to my membership on the steering committee (which, in addition to its "professional" members, also contained two representatives of FODOK and, separately, two representatives of implanted children), I could follow developments from close by. Twenty children from ages three to twelve years were implanted in 1993 and 1994, leaving twelve months for evaluative follow-up before the project ended. What did follow-up consist of? The plan was to go beyond measuring improvements in speech perception, and in two directions. Language development was going to be assessed, and a group of sign linguists was invited to collaborate with the project (though the collaboration would prove difficult). Secondly, they were going to look at quality of life outcomes. The questions addressed were "How does cochlear implantation aid the auditive functioning, the acquisition of language and speech, and the social-emotional development of implanted children" and "What are the specific possibilities and limitations of a pediatric implantation program in the Netherlands"? There was no control group. This had been considered inappropriate "on ethical and technical grounds." Preliminary applications on behalf of 130 children were received between September 1992 and September 1995, of which 106 were considered for the program. For the study, fifty-six of these were rejected, four had in the meantime been implanted, and twenty were to be implanted in the course of the new program. Another twenty-four were put onto a waiting list.

In 1994, the Utrecht group submitted its application for a pediatric program together with another school for deaf children (Effatha) that, unlike the oral IvD, used the mixture of spoken language and sign known as "total communication." In approving the Nijmegen project, the Health Insurance Council had again recommended that the two groups work together, and the Utrecht proposal explains that their project would permit data to be analyzed together. However, this project would differ from the other in a number of respects, including the inclusion of a control group. Moreover, the influence of the communication medium (total communication versus oral) on the results would be considered. The project would seek to answer the question "How are these children best accommodated in existing educational structures?" Should they be offered total communication–based schooling together with other deaf children, should they be placed with hard-of-hearing children, or would they benefit from regular (mainstream) education? Twenty children between three and eight years old would be implanted over a period of two years, leaving one to two years for follow-up and evaluation. The total group of forty that the two projects together would implant should be "representative of deaf children in the Netherlands" (though no reference at all was made to ethnic diversity or to the use of sign language or speech in children's family or educational environment).

The program application offers a clear picture of what implantation was expected to yield. "The question is, will the child take its place in the world of the deaf, or can it learn to hear, to lip-read and to speak so well that it can also take its place in the hearing society. That is what many parents wish for their child." Anticipating the results to be obtained, the proposal goes on, "Cochlear implantation offers, in principle, the possibility of integrating children in a regular school near to their homes. This gives parents more possibility of raising their children themselves." One year after implantation, an assessment would be made as to whether the child should remain in deaf education. "It is to be expected that, after implantation, the child should be regarded not as wholly deaf but as hard of hearing. Obviously its educational placement must then be adjusted. Somewhat later it is possible that some children will even be candidates for transfer to regular education." Surely this was strange. A hospital was applying to a fund for investigative medicine with mainstream educational placement as its main goal. What does this have to do with medicine? How could a group led by surgeons claim to be able to provide answers to questions of educational placement, and in the context of a medical protocol?

Taking Positions, 1994–1995

Educational psychologists in Sweden had studied the effects of their country's switch to sign language–based education on deaf children's cognitive and social development.[14] In various studies, groups of children had been followed over a number of years. Deaf children in the 1980s, brought up with sign language, did very much better than their predecessors on all kinds of tests of cognitive ability: ability to read and write Swedish, mathematics, and so on. What is more, the gap between hearing and deaf pupils had been reduced substantially. In the 1960s, many Swedish deaf children, like their peers in most countries, acquired only minimal reading and writing skills. By the 1980s, although deaf children needed more time to complete a test, the average reading ability was almost equivalent to that of hearing children. And about half of the deaf children did as well as the hearing children in mathematical tests, even without extra time.

"The average level of theoretical knowledge has risen dramatically," according to Kirstin Heiling, an educational psychologist and for a time advisor to one of the Swedish cochlear implant programs. Indeed, there was evidence that some of the deaf children, "whose hearing parents started to use signs when the children were two years old or younger, last spring performed just as well as hearing age-mates on a standardized reading achievement test."[15]

Surely this was what the objective had to be—not parroting speech but enabling deaf children to develop cognitively and socially, as normal well-integrated children. How could that best be achieved? I was convinced that Sweden had made more progress than any other country I knew about, certainly more than the Netherlands. And all this had been achieved through collaboration, not conflict. Swedes tend to attribute the acceptance of sign language in their country to an effective partnership between the Deaf community, the parents' organization, and scholars and scientists working on sign language. Partnership, consensus, and mutual respect are characteristics of the Dutch political culture, too.[16] I had seen for myself that in Sweden doctors leading cochlear implant programs had had to take the views of the Deaf community seriously. Surely this mutual respect would make it much easier for parents to make well-considered decisions regarding the cochlear implant? Dissent and conflicting opinions made matters much more difficult for them, and yet almost everything written about cochlear implantation was in support of one view or the other.

The fact that little or no debate had taken place in the Netherlands did not, in my view, imply that the implant teams could be left to make policy alone. The Dutch Deaf community had to formulate its own views, and debate had to follow. Consensus regarding the proper use of cochlear implants was the stage after that. By 1994, I was beginning to see that cochlear implants could, as had happened in Sweden, be used to complement sign language rather than to undermine or discredit it. (Not that there was no disagreement in Sweden. There certainly was. But what was at stake was different.) In just about every document on cochlear implants I had read, the word "language" was used as a synonym for "spoken language." Phrases such as "The implant would give children access to language" implied sign languages did not exist, or were not worth mentioning. There was a long way to go. Clinicians would have to learn to accept the equal linguistic standing of sign language. They would have to understand its importance for the emotional, cognitive, and social development of deaf children. They would have to respect the legitimate aspirations of the community of sign language users. The Deaf community would also have to respect parents' needs, and rights, to decide what action is in the best interests of their child. It would have to learn to help parents understand what growing up as a member of the community of sign language users could mean. It would have to make itself accessible. Anthropologists Gary Downey and Joe Dumit had written of "Getting each side to recognize and accept the legitimacy of the other" as a research strategy.[17] This was coming to be my strategy too, sharpening the focus of my involvement.[18]

It had all kinds of implications for what I had to do. In the steering committee, where I now represented FODOK, I was starting to establish my role. Endlessly pointing out when the word "language" had to be replaced by "spoken language" was a simple example of what I found myself doing. As I had discovered at that first discussion with FODOK, the question of how the benefits of the implant are to be understood is a crucial one. There, I had questioned Jan's interpretation in terms of ease of family life. Now I was questioning audiologists' interpretation in terms of measured word comprehension. It had to be much more comprehensive than that. It had to be established in terms of the kinds of things those Swedish psychologists had been studying. But critique of assessment studies was not enough. I had to try to highlight the ways in which the technology was being used to privilege the one understanding against the other. For example, the implant professional encourages the parent's hope that with the aid of the implant, the child can be "restored" to the hearing world. "If you want an implant, you have to help your child make the most of it. To do that you must stick to spoken language. Avoid signing." In justifying the implant politically, economies in the costs of education were promised. If implanted deaf children could be moved into much less expensive mainstream education (as the Utrecht proposal anticipated), a good deal of money could be saved. It could be shown to be cost-effective.

By 1995, studying the development of cochlear implantation was becoming a very different experience. It was not only because I had come to know many of the participants personally. More importantly, positions were becoming fixed. To talk of opposing sides having emerged would be too simple an interpretation of what was going on. The implantation teams were beginning to notice that opposition to their work, controversy, was emerging. A leading figure in the Dutch Deaf community, himself a doctor, wrote in 1994, "I am no absolute opponent of CI. I think that it does have value for the late deafened . . . I am in principle absolutely opposed to the implantation of prelingually deaf children . . . For postlingually deaf children, perhaps the possibility of an implant must be retained where the wish is very strong. But then only accompanied by professional support that enables the child to retain his/her Deaf identity."[19]

As opposition from the Deaf community appeared to be growing, the implantation teams began to feel the need of some channel of communication.

There were no contacts with the leaders of the organized Deaf community. I noted my experiences at one event:

> Friday 27 May 1994. Montpellier, France. The 2nd European Symposium on Pediatric Cochlear Implantation. Peering out at the warm sunshine and enticing terraces of southern France, participants are also a little apprehensive. There are rumors of a massive protest by the French Deaf community. Nevertheless, sun and wine are irresistible. I sit in the sun chatting with a leading implant surgeon and with one of his colleagues from rehabilitation: a young woman with whom I have become friendly. From the conversation it becomes apparent what is at stake. An application for the funding of a further series of pediatric implantations is being prepared. The Dutch groups are going to offer to host the 4th Symposium. Perhaps I'd like to organize a session on social and ethical issues . . . ? Anyway it would be disastrous for patients too if work had to stop. And it would be disastrous if controversy on the scale or of the intensity of what had happened in France or Germany were to come to the Netherlands. Major opposition from the Deaf community, or from FODOK, could seriously endanger the funding application. Was dialogue possible? Might FODOK be willing to support an application for funding? Although I couldn't speak for the Deaf community, I explain that support from FODOK was not inconceivable—provided their major concerns could be met. Among these, I thought, provision for a long-term study of the effects of being implanted on socio-psychological development was preeminent.

In December 1994, the report of the French Committee on Medical Ethics (CCNE) appeared.[20] Having obtained the report from French colleagues, I set about making it known in the Netherlands. A summary, in Dutch, was published in the national deaf periodical *Woord en Gebaar* (Word and Sign) in January 1995.[21] My short piece elicited an enthusiastic reaction from readers and considerable interest on the part of the implant teams. I was convinced that a set of practices and a set of principles acceptable to all sides were in reach. In a special issue two months later of *Woord en Gebaar* devoted to implantation, I returned to the same theme. Referring once more to the French document, I wrote that "their position must also play a role in the discussion now beginning in the Netherlands over the future financing of CI. It is a position that is sufficiently sensitive to the different positions and makes possible a dialogue between deaf, parents, and professionals. The implant teams (at least in the Netherlands) are now ready to engage in such a dialogue."[22]

My optimism seemed to be born out. In a talk at a Dutch conference in May 1995, a member of one of the implant teams said, "I'm asking for openness from the sides both of the CI teams and of the deaf community, for respect for each other's perspective, and for dialogue. Professor Blume has already said it in his article in the March issue of *Woord en Gebaar*: the teams engaged in implantation of deaf children in the Netherlands are open to dialogue with representatives of the deaf community."[23]

By mid-1995, it was becoming clear to FODOK that they needed a position of their own. In the brief document that was produced, the organization stressed that until it had become clear what the implications of the implant were for the social-emotional development of the child, it would not take a view for or against implantation. Until that time, the view was that current projects should be continued, but coupled with long-term research designed to provide the missing insights. Parents contemplating implants would have to be better informed, including about the Deaf world and the possible value of sign language. These are my notes from such a discussion:

> Friday 27 October 1995. Uddel, the Netherlands. 170 people have gathered for a two-day workshop on cochlear implantation. They include representatives of the deaf community, members of one of the implantation teams, and representatives of FODOK, as well as many members of all constituencies. There is a sense of this being an historic occasion: perhaps the first such gathering in Europe. After a day devoted to formal presentations, the second day is given over to discussion. Although the busy surgeons have left, many leading figures from all constituencies have stayed for the working groups. Everyone is a little surprised at the extent of agreement and the wish for dialogue. The Deaf people don't seem to have any objection to CI provided it is made compatible with sign language use. Members of the implant team explain that they would be perfectly willing to collaborate in the long-term evaluation that FODOK wants. Everyone agrees that much is as yet unknown. In the final plenary session, FODOK chair proposes that all organizations represented should together sign a letter to the Health Insurance Council, asking them to extend the current project for 5 more years, and set up a study that would focus on all the issues concerning the deaf community. Her proposal is put to the vote. Slowly, with more or less hesitation, almost every hand goes up. Afterwards a deaf man whom I know tells me that he'd come full of aggression . . . but it had gone!

By the end of 1995, some harmonization of objectives and projects seemed to have come about. The implant teams were concerned principally with finding the funding they needed to work through their waiting lists. They realized that public controversy could only reduce the chances of the procedure being funded. Dialogue would help. FODOK wanted to see parents "empowered":

enabled to decide in an informed way. They wanted the technology to be more thoroughly evaluated in terms of the things that were really important. The Deaf community, perhaps pleased at being accepted as a legitimate partner, wanted above all to consolidate progress in the acceptance of sign language. They seemed willing not to oppose implantation provided it did not threaten progress in this direction. The consensus reached in the course of 1995 was marked by the establishment of a forum ("platform") in which dialogue between the principal interest groups would be institutionalized. Under the auspices of this platform, which had an independent chair, work started on the drafting of a collective letter to the Health Insurance Council.

Consensus under Pressure, 1996–1998

Precisely at this time, late 1995, the report to the Fund for Investigative Medicine on the Nijmegen/IvD pediatric implantation project was nearing completion. For the implantation teams, this report was crucial. It had to convince the Health Insurance Council that the benefits of cochlear implantation in children had been proven. The Council would then recommend that the minister of health agree to reimburse pediatric implantation from health insurance funds. Even within the teams, differences of opinion emerged regarding what should or should not be in the final report. Parts of the study were being done by research groups independent of the implant team: by a group of sign linguists (who were focusing on the interrelations between the use of sign and spoken language by implantees) and by a medical technology assessment group (that was looking into the broader social and ethical implications of the procedure). The conclusions they were reaching, emphasizing all that was not known, the gaps in knowledge, and the assumptions on which the analysis was based, did not fit easily with the general conclusions the implant team wanted to draw. As a member of the steering committee, I was privy to some of these discussions. I could see the tension between those who wanted to restrict the report to hard quantitative data (largely limited to measured gains in speech production and perception) and those who wanted to refer to gaps in knowledge and the controversy surrounding the technology. What mattered above all was how the report would be received. Disagreement within the team was resolved largely in private. The report, with only brief reference to gaps and controversy, was submitted in early 1996, and the waiting began.[24]

More than 70 of the report's 180 pages were devoted to the results obtained. They were complex. For thirteen of the twenty children, spontaneous language production was analyzed. How did an implanted child choose to communicate in a familiar setting? For the study, the children communicated with a family member (often the mother) at home, and these interactions were video recorded and analyzed. Many of the children used a mixture of spoken Dutch, sign-supported Dutch,[25] and sign language utterances. The analysis showed a slight increase in use of spoken language and a slight decrease in the use of sign language over time. The researchers also studied whether the children's spontaneous speech could be more easily understood as time passed. Here, contrary to expectation, no change was found. Only two and a half pages were devoted to the children's socioemotional development, based on questionnaire responses of their parents. Implanted children were reported to be more stable and more natural in their behavior than non-implanted children—both before and after implantation. A quality of life survey, again completed by parents, showed no significant difference between implanted and non-implanted children. Children with an implant did seem to be doing better at school, though the report acknowledged that significantly more resources (extra speech therapy, extra supervision) were invested in these children. Moreover, they had been doing better than average even before implantation. Though there was definite improvement in the children's ability to interpret spoken language, no clear gains on any other dimensions could be demonstrated. A problematic result, this was attributed to the duration of the study: "too short to permit detailed evaluation of developments in relation to language or socio-emotional development."

In a few pages, towards the end, the report acknowledged that there had been opposition to the implantation of deaf children. This section, headed "Ethical Aspects," noted that the primary responsibility lies with parents: a responsibility that they prefer to share with doctors, paramedics, psychologists, and teachers, although "information from the deaf community is also of great importance." The Uddel conference was noted as beginning a dialogue that should bring the two sides together. Dialogue is important and the platform can play an important role. "What is not acceptable is that the deaf community forces its views on, or brings emotional pressure to bear on parents by presenting implantation in a bad light, let alone that it brings pressure to bear on those who must decide over the financing of cochlear implantation."[26] This sentence merits some reflection, and I shall return to it later.

In February 1996, the report was made public. Perhaps I should not have been surprised at the kind of article that followed in one of the country's leading newspapers.[27] But I was. All those familiar refrains—here they were again: "the wonder of hearing sound again" . . ."the incomprehensible objections of the deaf" . . ."the probability that a deaf child will be able to integrate in regular education." "You make a deaf person into a very hard of hearing one" was how one leading surgeon summarized the improvements. Another leading surgeon was quoted as saying, "With this technique you can reduce the consequences of deafness for almost all Dutch children. And that for five million guilders per year, including after-care." It was not too difficult to see how this article had come to be written.

The controversy, so clearly influenced by the political agenda, made me endlessly suspicious. When, in September 1996, I was asked to join the program committee of the fourth European Symposium on Pediatric Cochlear Implantation, I hesitated. Was I being asked to show participants the value of dialogue, to show how far we had come in the Netherlands? Was this a valuable opportunity that I could use? Or was I myself to be used in some strategy of which I was not fully aware? I had learned that a willingness to discuss in a respectful and civilized manner where nothing was immediately at stake did not necessarily translate into political compromise.

As political decision making appeared to be approaching, the modus vivendi that had emerged among platform members began to break down. When the long-awaited response of the Health Insurance Council—in the form of advice to the health minister—finally became available, matters deteriorated still further. It had taken twelve months since the report had been submitted to the Fund for Investigative Medicine. Now the Health Insurance Council was advising the minister to reimburse implantation. All doubts seem magically to have been resolved: "Children with acquired deafness reduced their delay in linguistic competence. For deaf born children the results were less clear. . . . The quality of life of children with an implant seems 'overall' to be somewhat better than children without."[28] The fact that dialogue had taken place over the previous twelve months was now being used as evidence of responsible decision making in the Netherlands (compared to other countries), as evidence that "the deaf world is not opposed to the application of CI by children, but regards it as experimental, until long-term results are available." Encouraged, the implant teams were now unwilling to see a definitive decision postponed until a long-term evaluative study (bringing in social, psychological, and cognitive factors) had been conducted. The apparent consensus reached at Uddel and its institutionalization in the platform had been overtaken by events.

As far as I was concerned, neither the project report nor anything that had happened since it had been written justified these conclusions. And I was unhappy at the way in which the work of the platform was being used as evidence of tacit support. Informal inquiry suggested that acceptance of the council's advice by the health minister was more or less automatic and imminent. Any response from "consumers" would have to be rapid and was unlikely to have any effect. But the Deaf community responded in sufficiently strong terms that the minister decided to make time for consultation with all interested parties. These consultations took place soon afterwards. FODOK, too, was invited for a discussion with the civil servant responsible for preparing the minister's reaction. This discussion, in which I participated, was a tough one. Something curious became apparent.

Early in 1996, the government departments of Health, Welfare, and Sport (WVS), and Education, Culture, and Science (OCW) had, jointly, appointed a committee to develop proposals for how official recognition of Dutch sign language could best be given form. This official committee had been asked not *if* it should be recognized but *how*. The Netherlands was apparently moving in the direction that Sweden had taken fifteen years before. The committee, which was chaired by Anne Baker (Tevoort's British-born successor as professor of linguistics at the University of Amsterdam) would look into the implications for education, rights and legal process, social work, and so on. At the time the discussions regarding the implant took place at the ministry of health, this committee had not yet reported (it would do so in the course of 1997).[29] But the point is that officials of the ministry of health had seen no connection between routine provision of cochlear implantation and the government's commitment to the official recognition of sign language. So far as they were concerned, cochlear implantation was just one more complex and expensive medical intervention, the evidence for which was not clear-cut. Getting them to see a connection was not easy. Administrative organization stood in the way. Sign language recognition had to do with the handicapped, and that was a different part of the ministry. The official we spoke to said that he expected the minister to reach her decision within two months. At a platform meeting in this period, it seemed that bitterness and frustration was now so great that dialogue was no longer possible. The platform's independent chair doubted whether it made any sense to continue to meet.

The minister could not reach a decision in two months, and the delay did not help. By the subsequent meeting of the platform (in September 1997), relations were becoming even more strained. None of the deaf members showed up. Representatives of the implant teams who had once seemed so open and conciliatory no longer were. For them, I now seemed to represent the (absent) Deaf community. Someone said, "If you put the effort needed into doing your best for implanted children, providing them with the oral environment you needed, you couldn't teach them sign language." Everyone had previously been willing at least to pay lip service to the idea that you could combine the benefits of the implant with those of sign language. Now, under pressure, it seemed that they were retreating to older views.

In November 1997, the Dutch minister of health reached a decision. To the astonishment of many, she postponed admitting pediatric cochlear implantation to regular health insurance funding, regarding it as potentially worthwhile but not yet adequately proven. The minister, herself a physician, wrote that the effectiveness of the use of the implant with children had not yet been unambiguously established. There were, to be sure, indications that this form of care could be of value for the (long-term) development of some deaf children. Implantation had to be continued, but under experimental conditions, according to strictly defined protocols. The effectiveness of the procedure in the long term had to be established. The minister's preference was that pediatric implantation continue to be financed from the Fund for Investigative Medicine. The Health Insurance Council, under whose auspices that fund operated, disagreed. They stood by their advice.

So far as the platform was concerned, the resulting impasse, with ministry and council disagreeing, was problematic. Formally the platform, like the minister, had taken the view that a study of the long-term social and psychological effects of the implant was desirable. But if that argument was now being used to hold up funding, it was another matter entirely! Where was the money for further operations to be found? The implant centers faced a real problem, as did the growing number of parents whose children were now on a waiting list. If age of implantation made a real difference to the results, then for a child to have to wait two to three years could be disastrous. It was in everyone's interests that the unclear situation be resolved. The platform would write a collective letter, pointing out that the existing situation was undesirable and asking that funding be made available for an extension of the previous experiment. All members of the platform agreed to this.

Parents needed to know where they stood. The implant centers said that those who could afford it were offering to pay themselves or were sending their child abroad for implantation. Since it was unclear who would then be responsible for their rehabilitation, or how this would be paid, this was an undesirable state of affairs. In the course of 1998, the situation became more complex as behind-the-scenes lobbying intensified. It became clear that the Health Insurance Council was unwilling to finance an expensive long-term study of implant recipients, being convinced that it would add little of importance. Whatever the health minister thought, in their view enough was known of the benefits of the implant. They were, however, willing to provide interim funding, to keep the programs running till 2000. The minister, we understood, was hoping that some consensus could be found.

Naturalization of the Implant

By early 1999, it looked like the platform would soon cease to exist. Not that it cost much to run, but there were now no resources at all, and the ministry of health said it could not help. Then came a new suggestion. Although it could not finance the platform as such, the ministry would be very interested in a collectively produced clinical guideline. Perhaps some funds could be found for an exercise like that. The nonmedical members of the platform, myself included, were unsure as to what "clinical guidelines" looked like. Still, trying to reach agreement on rules to govern implantation practices seemed a worthwhile exercise. But there was not much time. Interim financing of implantation was available only until January 1, 2000. It would be best if implantation could be continued after that date on the basis of clinical guidelines that enjoyed broad support. For that, the work would have to be completed by the summer of 1999. And so, in January of that year the cochlear implantation platform set out on what was to prove its terminal adventure.

The drafting was delegated to a working group, which speedily set to work. Eventually I too got the opportunity of commenting on the drafts they were producing. I had heard rumors that neither the organization of the deaf (Dovenschap) nor FODOK were entirely happy with the way things were being pushed along. But the work was surely important, and I remained committed to the idea of guidelines to which we could all subscribe. But reading the draft, it became clear that there was still a problem. What was being produced was a *clinical* guideline, and I simply could not fit my concerns into that format. At the end of July, I wrote to the drafting group:

> The Deaf community, and others who worry about these things, are being asked to abandon their objections to CI in exchange for (more?) careful and responsible practices on the part of the implant teams, whilst the major difficulties lie outside the realm of that practice. If, as is possible, large numbers of parents of implanted children choose to move their children into regular education, and numbers in deaf schools become so small that those schools (and hence bilingual education) become non-viable, what then? If deaf children are increasingly educated in regular education from an early age—which CI seems to promise—and as a result have little or no contact with other deaf children, what then? Are we not then defeating some of our fundamental "objectives"? CI, certainly if offered on a large scale, has implications which are not easy to incorporate in clinical guidelines—respecting as always the "patient's" right to choose—but with fundamental implications for the realization of general objectives.

I never discovered what happened with the draft. No version of the guidelines ever came to the full platform for formal approval. We failed. But it began to seem that this was of no consequence. The approach of the new millennium and the end of the interim financial arrangements had led to even more intensive lobbying by the hospitals. The minister could not or would not wait any longer. Ministry officials had, once more, requested a meeting with both FODOK and Dovenschap. This time I was not invited to participate. Both organizations had new leadership and they were well able to plead their cases without my help. Except that this time dialogue was not what the officials had in mind. They simply had a statement to make.

In November1999 Paul van den Broek, pioneer in the introduction of cochlear implantation to the Netherlands, delivered the public lecture with which a full professor in the Netherlands marks his or her retirement. The hall was filled to capacity. Lots of speeches were made. I had been asked to write something about the event for *Woord en Gebaar*, and on the way back to Amsterdam I began to formulate what I would report.

> Hundreds of people came to listen to Professor van den Broek, who has been a driving force behind the introduction of cochlear implantation in the Netherlands. As I listened to his lecture, I had to think back to the many discussions we'd had since I started studying the controversy around cochlear implants more than seven years ago. In the beginning, we had disagreed about almost everything. When the newspapers or the television devoted any attention to CI, van den Broek was always quoted. Somehow or other those reports always seemed to have headlines like "Cochlear Implants Better Than Sign Language." But this lecture was different. Much of it covered the long history of the dispute over whether deaf education should make use of sign language or spoken language, and the notorious Congress of Milan, which marked the decline of sign language–based education, more than a century ago. Needless to say much of the lecture was about cochlear implants and their value for deaf children. Seven years ago, I don't think van den Broek would have told his audience that sign language was valuable for young deaf children. I don't think he would have bothered to say that a child could have an implant and still be bilingual. I don't think he would have pointed out how little was known about the psychological or social effects of implantation in children. Now he did all of those things. Over seven years, the differences between us have narrowed enormously.
>
> He had planned to complain to his audience that Minister Borst *still* hadn't approved implants in children for reimbursement, making the Netherlands the only country in Western Europe where the costs weren't covered by health insurance. But on this point he had to change his text, because that had changed just a few days previously . . . Two years of discussion in and around the Ministry of VWS and, I suspect, some pretty intensive lobbying in the past weeks, had finally produced a decision.

> What does this mean for the Deaf community? In my view, a new, and very different challenge. It is no longer a question of trying to stop the implantation of deaf children. That is now going to happen on a larger scale than in the past. In years to come, many deaf children will have implants: maybe as many as half. A few years ago the French National Committee on Medical Ethics reviewed what was known about CI and came to the conclusion that all deaf children needed to start early with sign language. This was true even if they might later be candidates for a cochlear implant. They were right. I am convinced that deaf children using cochlear implants will need sign language. This is a question of psychology not audiology. Their basic education needs to be bilingual (in NGT and in Dutch), so that they too will become proficient in both languages. One challenge for the Deaf community, and the schools for the deaf, will be to get this message across. Many parents will probably hope to see their implanted children move quickly into regular education. Some may even want to think that, thanks to the CI, the child isn't deaf any more. I hope we can now count on the implantation teams not to encourage them in this view. But still, there is a challenge. How can the new parents of deaf children be shown that rapid transfer to regular education isn't necessarily the best thing for the child? It will often be best to wait a few years, until the child has mastered sign language. I personally see a much greater role for the family support services and for the schools for the deaf in discussing with parents what is best for their particular child. It is here that parents should get advice regarding implants and other aids. Parents need to be able to discuss what a CI would mean for the child's schooling, what it would mean for communication in the home. They need advice from people who aren't committed to any particular technique but who have seen many deaf children grow up. This is the way I'd like to see it going, rather than that parents directly approach a hospital department of ENT. We have to recognize that the surgery isn't the most important thing about cochlear implantation. And we have to recognize that even with an implant, all those difficult questions about language, culture, and identity are still there. I'm sure Paul van den Broek would willingly admit that surgeons aren't expert in these areas. I hope his successor will too!

With the decision to reimburse, the steam went out of the platform and the goal of producing consensual guidelines was quietly forgotten. It was becoming clear that large numbers of deaf children in the Netherlands would be implanted. What still had to be discussed was how the implant would be used. It was all the more necessary to ensure that it would not fuel the rebirth of oralism. That need not be so, but all kinds of policies, notably on mainstreaming, could lead that way. The official recognition of sign language had still not taken place. But how to keep the discussion going? It looked as though the opportunities for intervening, and specifically for making the Swedish way of doing things relevant for the Dutch situation, were exhausted.

Valuable opportunities for saying the things I felt needed to be said did occasionally arise. An invitation to comment on the Dutch reaction to a draft report by Swedish psychology professor Gunilla Preisler to the Council of Europe, in late 1999, was one.[30] Hopefully Preisler would be tying the issues raised by the implant to the sorts of issues—such as minority rights—with which the Council of Europe had become particularly associated. Here is part of what I wrote:

> The Schools for the Deaf are more or less committed to bilingualism . . . Steps towards the official recognition of Dutch Sign Language have been taken. This is the public agenda, and it would imply that here, as in Sweden, little of the literature on cochlear implants is of relevance. Like Sweden, we would logically have to assess the social costs and benefits of CI on the basis of the ultimate social goal to which the implant is only a means. Collective commitment to bilingualism means that the value of the implant, including of its specific use in the Dutch context, has to be assessed in relation to this goal rather than in relation to the implicitly assumed goal of oralisation.
>
> Privately, rather than collectively, more and more parents of young deaf children seem to see CI as a desirable alternative to learning sign language. In Sweden functional sign language communication in a family is a requirement for admission to a CI program. We are unlikely to go so far. But if we are committed to bilingualism, how is it possible that "new" parents come to this conclusion? How is it possible that they see CI as opening so easy a route to spoken language that sign language becomes unnecessary for them? Where does this message come from? Since the professional community of medical and other "helpers" is publicly committed to bilingualism, we must assume that they are not its source. From documentaries such as *Sound and Fury* recently shown on Dutch television, but reflecting the American situation?[31] The message conveyed by this documentary is one that all responsible organizations in the Netherlands, including most importantly the educational organizations, claim to have abandoned. But for the parents of children only recently diagnosed as deaf, the notion that a child can no longer be deaf will always be powerful and seductive. Such emotions can easily be mobilized for political purposes. The history of new medical technologies is full of examples of just this being done. But who would want to mobilize hopes in this way, today?[32]

In April 2000, it was announced that, despite having already taken a decision about the finance of implant procedures, the minister of health had nevertheless requested the Health Council (not to be confused with the Health Insurance Council) to report to her on current and anticipated developments around the technology.[33] I was invited to become a member of the committee charged with preparing this report. Paul van den Broek and Egbert Huizing, both now retired, would also be members. The committee's deliberations made clear to me that in a consensual political culture such as that of the Netherlands, compromises must be made but can also be required. We worked fast and efficiently. In August 2001, the chairman of the Health Council submitted the committee's report to the minister. Many of its recommendations concerned the future scale and optimum organization of implantation programs.[34] It was becoming clear that almost all university hospitals were planning to establish pediatric implantation programs, a dilution of expertise that the committee felt undesirable. But the report also referred to all that is not known regarding the effects of implantation on children's development. It referred to the possibility of prior sign language communication being a precondition for effective acquisition of spoken language with the aid of the implant. It referred approvingly to the fact that a "respectful dialogue between the CI centers and the organizations representing the deaf and the parents of deaf children has been established . . . which has prevented the kind of strong polarization that is found elsewhere." Arguing that the cochlear implant enhanced access to spoken language while accepting that many deaf children would have already started using sign language would make it much more acceptable to the Deaf community. That would reduce the chance of controversy. Next to "clinical guidelines regarding the appropriate application of CI, there is also a need of 'social guidelines,' covering the broad areas of education, care and the social integration of deaf people. A strengthening of the position of deaf people in society is important for their acceptance of CI."

In May 2003, the minister of health circulated a draft policy document for which the Health Council report provided the basis. In the introduction, the minister wrote that pediatric implantation was to be brought under an article in the law that allowed central regulation of selected medical interventions. Only "a small number of designated academic hospitals" would be permitted to carry out pediatric implantation. (Implantation of adults was not to be regulated in this way). For the four years for which the policy initiative was valid, the minister would designate four hospitals as authorized to conduct pediatric implantation.

This policy conflicted with the wish of each of the country's eight university hospitals to start doing pediatric implantation. In fact, each of them had already started, thus challenging the minister with a fait accompli. A year later, in the definitive version of the document that was sent to the Dutch parliament, this concentration policy had been abandoned under pressure from the university hospitals. The change of mind was justified on the grounds that "at this moment all university hospitals are doing pediatric cochlear implantation, and that all of them possess the knowledge, the expertise and the infrastructure needed to carry out the intervention in a responsible manner. Designation of just a few hospitals would thus lead to loss of experience and waste of capital already invested." All that the minister now required was that each center commit itself to national guidelines that they were themselves to develop. They had already begun work on these guidelines.

Pediatric implantation had now become a standard procedure in the Netherlands, reimbursed by health insurance. Today the majority of young deaf children are fitted with implants, and with the extension of neonatal screening for hearing loss, the age of implantation is decreasing. The surgeon who, years before, had told me that in time all deaf children would have an implant seems to have been proven correct.

Cochlear Implantation and Consensus Politics

Chapter 4 examined the spread of pediatric implantation from country to country, despite the criticisms of Deaf community leaders and advocates for Deaf culture. I argued there that we could understand this in terms of two sorts of differences: first, in the channels through which the technology and the discourse supporting it on the one hand and the critical discourse on the other flowed from place to place; second, in the processes through which each was "put to work" in the specific national context. In this chapter, I set out to examine these processes in greater detail.

What this ethnography of cochlear implantation in the Netherlands shows most clearly is the crucial role of the established decision-making structures under the aegis of which implantation took place and through which its future was decided. These decision-making structures were dominated by medical professionals, medical ways of thinking, and medical values. The initial program of pediatric implantation was funded by a Fund for Investigative Medicine, itself responsible to the Health Insurance Council. With memberships largely composed of health professionals, these organizations were accustomed—like equivalent organizations in other countries—to focus on proofs of safety and efficacy. Efficacy is taken to mean benefit to the individual patient, while "benefit" is interpreted in terms of the specific clinical variable, or end point, with which the relevant medical specialty is accustomed to working. So it was here. Despite the references to deaf children's quality of life, to their cognitive, linguistic, and social development, and to educational placement, the benefit of the implant was initially assessed by means of audiological tests. It was only in relation to speech perception that hard evidence for the effectiveness of the implant could be gathered (at least, within the time frame of the projects and with the skills available). I referred earlier to the tendency of the mass media to oversimplify in reporting the results of scientific investigations. This is not a tendency to which the mass media alone are prone. When, in 1996, the Health Insurance Council advised the minister of health to approve structural reimbursement of pediatric implants, it did the same thing, supporting its recommendation with a summary of what studies had shown that went far beyond what had actually been demonstrated.

Through structures like this and through many other formal and informal channels, the medical profession had opportunities for influencing senior decision makers that the Deaf community did not have. Earlier, I quoted a sentence from the report on which the Health Insurance Council based its advice in 1996: "What is not acceptable is that the deaf community forces its views on, or brings emotional pressure to bear on parents by presenting implantation in a bad light, let alone that it brings pressure to bear on those who must decide over the financing of cochlear implantation." This sentence is saying, in effect, that our perspective is the right and the natural one. To offer an alternative is in some sense immoral. It is also saying—though this becomes more apparent in the light of subsequent events—leave the lobbying, the exercise of pressure, to us. And this of course is what happened. Physicians and their hospitals' PR departments not only made effective use of the mass media, they also lobbied so effectively that the minister of health had little choice but to accede to all of their demands.

This is not to say that opportunities for the expression of alternative views were not created. Despite the lethargic response of the Dutch Deaf community, there were opportunities to speak, a reflection of the consensual political culture for which the Netherlands is known. The best example of this was the Platform Cochleaire Implantatie bij Kinderen (Platform Pediatric Cochlear Implantation), established in 1995, chaired by someone with no personal involvement in the issue. All parties agreed to be represented on it. The platform provided ample opportunities to the Deaf community to express its concerns and demands. Subsequent events, however, throw a different light on this institutionalization of dialogue. The functioning of the platform, indeed its existence, proved fragile and was threatened by the report of the Health Insurance Council that appeared to support uncritically the medical view. With structural financial provision seemingly in the offing, implant teams' willingness to engage in dialogue, to seek compromise, declined. And when financial provision had been secured, the implant groups saw little purpose in continuing the platform. They let it, and the half-developed clinical guidelines, fade into oblivion.

An irony was that despite (or perhaps thanks to) its inability to influence the course of events, the very existence of the platform could be used politically as indicating the lack of opposition on the part of the Deaf community, as it was in the Nijmegen/IvD team's 1996 Report to the Fund for Investigative Medicine. Its existence was used again by Dutch officials when responding to the Council of Europe in 1999 (even though the platform was by then in process of going out of existence).

Dissenting views, the anxieties of the Deaf community, the lack of evidence on the long-term or socio-cognitive effects of implantation—all had opportunity of expression in the Netherlands. The Report of the Health Council, with its references to the need for "social guidelines" and to the importance of "strengthening the position of deaf people in society" attests to this. Yet it made little difference. Having initially accepted some at least of the arguments of this Report, within a year the minister had been forced to backtrack by pressure from the university hospitals.

A similar fate had befallen the report of the French Committee on Medical Ethics on which, a few years earlier, I had pinned so much hope. Why do carefully considered reports like these have so little influence on actual practice? As discussed in chapter 1, the bioethicist William Gardner posed a similar question in relation to genetic enhancement.[35] His argument was that even were ethicists to agree that it was appropriate to ban enhancements that parents thought desirable, such a ban would fail in practice because the incentives to find ways around it would be too great. Gardner wrote, "Ethicists who hope to prohibit enhancement must carefully consider what legal and political mechanisms, national and international, will be sufficient to enforce that prohibition."[36] So far as cochlear implantation is concerned, those mechanisms were not, and are not, in place. Not only was the Dutch Deaf community unable to gain much hearing for its views directly, but even an authoritative body like the Health Council (or the French CCNE), having accepted some of their concerns as legitimate, could not make its arguments stick. In the political arena in which decisions were taken, professional interests and the demands from growing numbers of parents of deaf children proved far more powerful.

Chapter 6

Contexts of Uncertainty

Parental Decision Making

Ever since the early 1980s, parents of deaf children have been central to the cochlear implant controversy. At that time, when implanting children was seen as a risky step that few surgeons were willing to take, organizations of parents of deaf children were angry and anxious. The French organization, ANPEDA, considered the step premature. Its leaders resented the idea of their children being used as experimental guinea pigs. The British organization, NDCS, was worried at the unrealistic expectations on the part of parents to which exaggerated publicity would lead. Nevertheless, in the course of the decade it became clear that this was the direction in which both the practice and the technology were evolving. Manufacturers set about developing smaller flexible devices that could accommodate children's growth. Discarding the lower age limit would greatly increase the potential market for the device. Clinicians came to believe that the chances of success, viewed in terms of enhanced speech perception, were greater in children. When the Food and Drug Administration approved use of the Nucleus device for children in 1990, it paved the way for a rapid expansion in the scale of pediatric implantation. Though they continued to urge caution and insisted on further research into the long-term effects of implantation, the opposition of parents' organizations faded. Many of the newly established implant centers, including those in the Netherlands, were particularly keen to work with children. The nature of parents' involvement in the politics of implantation gradually changed. By the 1990s, it was less the positions their organizations adopted that mattered than the aggregate decisions, for or against the implant, that individual parents took in private. Not adult deaf people but the parents of deaf children, 90 percent of whom are hearing, were coming to be seen as the principal "consumers" of the technology. From the point of view of the implant manufacturers, the parents' decisions, in aggregate, would determine the commercial success of the technology. From the point of view of the Deaf community, their decisions would determine the future of the community.

Given how much was at stake, it is little wonder that the controversy came to encompass questions of parents' rights, responsibilities, and expectations. For the medical professionals who involved themselves in the debate, parents' rights were clear and their competences not at issue. From their point of view, parents have the right and the responsibility to decide on behalf of their children, however difficult. Nothing must be allowed to stand in the way of parents making up their minds freely and autonomously. When the Deaf community argued that parents have an obligation to listen to them, opponents appealed to the ethical principle of autonomy. The Deaf community could not be allowed to interfere with parents' right to exercise free and autonomous choice: a right that did not, however, preclude advice or influence from the medical profession itself. Critics questioned the competence of hearing parents to decide in their children's best interests. These parents know nothing of the lives of deaf people, and for them deafness means only tragedy and isolation. Their emotions, their desperate hope that their child be made "normal," like themselves, stands in the way of their understanding or acting in the child's best interests.

Bonnie Poitras Tucker, a deaf law professor at Arizona State University, developed a different argument in favor of implantation.[1] It was that disabled people, or the parents of disabled children, are under an obligation to take all possible steps to ameliorate the effects of the disability in question. If they fail to do that, they have no moral right—and might ultimately have no legal right—to benefit from collective provision for the disabled. Resources, and hence provision, are limited. Tucker wanted to make claims on resources dependent on "responsible action," so that "an individual who chooses not to correct his or her deafness (or the deafness of his or her child) will lack the moral right to demand that others pay for costly accommodations to compensate for the lack of hearing of that individual (or his or her child)."[2] If cochlear implants work, and as far as Tucker was concerned they did, then deaf people (or parents) were under an obligation to undergo implantation or else forego their rights to costly services or benefits provided by the state.

In the last few years, a number of empirical studies of parents' expectations regarding the implant and their experience of the decision-making process have been carried out. The picture that emerges from these studies is a surprisingly bland one, little marked by doubt or controversy. One study in the United States found that most parents expected their children's understanding of spoken language to improve after implantation and had no sense of possible problems.[3] A British study found that 38 percent of parents expected improved hearing, with 23 percent also referring to psychosocial benefits "such as reduced social isolation and improved behavior."[4] Another study in the United States found that "the single most important factor in the decision to have the child receive an implant was a desire on the parents' part to have a child who might function as a hearing person" (thirteen of thirty-five interviewed).[5] A fourth study found that parents' motivations varied. For some, it had been the possibility of oral communication that mattered most, even if that were at the cost of the child's making (deaf) friends. Others saw it in exactly opposite terms. For them, it was the child's social relationships that mattered most, even if that meant sacrificing easy communication with hearing people.[6] Significantly, these studies were all conducted under the auspices of an implantation center and included only parents whose children had previously been implanted there. A Japanese study of parents of deaf children living in the greater Tokyo area chose a different sample of parents. This study, unlike the others, included parents who had rejected implants for their children.[7] Here too, the most frequently mentioned expected benefit was improved auditory skills, with some expecting that their children's speech would also improve. But in this report, unlike the others, some parents "expressed concerns that . . . incomplete improvement in hearing and speech would leave children without a clear identity in either the hearing or deaf communities." These parents were worried that an implanted child would be neither hearing nor non-hearing, "a vague kind of way to exist." There is nothing in any of the other studies to suggest that the parents interviewed (or the researchers doing the interviewing, for that matter) had given much thought to the question of the child's identity, as hearing, deaf, or "something in between."

The British and the Japanese studies asked parents about their experience of making up their minds. Had they considered alternatives to implantation? How adequately had they felt themselves to have been informed? How troubling, how difficult, had they found the decision? The British study, funded by the Cochlear Corporation and conducted in Nottingham University's department of community health, found that the majority of British parents had found the decision over implantation relatively straightforward, as they believed their child had nothing to lose and everything to gain. There was "*really no alternative*" (italics in the original). Fifty-nine percent of parents "believed themselves to have been well-informed about the intervention, and that there was nothing of relevance which they had not considered before sanctioning implantation. 'The information we've had has been first class all along the way. We felt we were in very safe hands.'"[8] There is no mention in this British study of something that appeared as a major finding in the Japanese one: parents' doubts about acting as surrogate decision makers. The Japanese researchers, Suguro Okubo, Miyako Takahashi, and Ichiro Kai, found that "all participants who considered cochlear implantation for their children were concerned about not knowing the wishes of their child. Even when families felt that the benefits would exceed the risks, some participants hesitated to decide on behalf of their child until the child was able to voice a preference."[9] British parents, it seems, had no such anxieties.

At least as reflected in these studies (with the exception of that conducted in Japan), parents represent themselves as the "informed consumers" who figure so largely in current health policy discourse. Their expectations (mainly formulated in terms of improved speech perception) correspond to the most-studied aspect of implantation; they felt themselves to have been well informed and to have considered all relevant issues. The decision had not been difficult. How could this have been so? How could the decision have been so unproblematic for these British parents but so complex for their Japanese counterparts? How could a British parent have taken the view that there was "really no alternative"? What did it mean to be "well informed" so far as these parents were concerned?

Parents' expectations of the implant are not formed only, and perhaps not even principally, by the information they receive from an implantation center. Growing up in a culture that expects so much of medical technology, most American and European parents will have spent their lives exposed to the consistent message of hope reiterated by the mass media. So far as the cochlear implant is concerned, that message, often supported by personal accounts of successful implantation, has been a clear and simple one. Based on his professional experience in conducting preimplantation psychological assessments, Robert Pollard came to the conclusion that candidates and their families often "present with oversimplified or distorted perceptions about cochlear implants based on material they have been exposed to in the popular media."[10] He stressed the need for parents to understand not only the implant and the possibilities it offers but also the alternative: the possibilities of life without an implant. And while medical professionals could play an important role in providing comprehensive information, in helping parents make a reasoned and informed choice, that is rarely what they do. Looking at the information material available to medical professionals and on which they base their advice, Pollard concludes, "Knowing little else about the ordinary lives of deaf people, what parents would leave such a presentation with anything other than the message that, if they love their deaf child, they should pursue cochlear implantation in order to prevent such catastrophes."[11]

Early in my study, when pediatric implantation in the Netherlands was quite new, I spoke to the Dutch mother of a child who had become deaf from an early attack of meningitis and was then approaching four years of age. Alice was also a good friend of mine. She told me how she and Francis, her husband, been advised that their son might be an excellent candidate for an implant since he had virtually no residual hearing. They were told to come back and discuss the possibility more seriously when their son, Simon, was a little older. They put it out of their heads, having in the meantime decided to learn Dutch Sign Language. Alice explains the reaction of the specialist who had been seeing their son when she returned to tell him of their decision:

> He demanded an explanation . . . It was an unfair discussion. I referred to things . . . something by a child psychologist I'd read who explained how vitally important communication at a young age was for intellectual development. It appeared he was a child psychologist as well. Then you get all kinds of arguments based on authority . . . Looking back I don't think it was fair. It's a time at which you have to take such difficult decisions. He is totally against sign language. He told me that if deaf people came to him for examination with an interpreter, he insisted that the interpreter be sent away. "You can read lips. I talk clearly enough" . . . It is so difficult.

Alice's interview with this medical specialist was if not coercive then certainly neither empowering nor supportive. It is not what she had expected or hoped for.

Whatever ethicists may say, parents are not autonomous and are not seeking to free themselves from external influences in deciding what to do. They experience decision making as a heavy burden, and it is help and guidance that they seek. Perhaps "autonomy" is experienced more as "abandonment."[12] Thomas Balkany, a well-known American implant surgeon, observed that talking with other families, especially families with implanted deaf children, is likely to have far more influence on parents than the "scholarly debates."[13] Nor, despite the ethical principles, does real world medical practice assume that parents are autonomous in their decision making. Rather it depends upon the crafting of an appropriate network of influences and relations of trust. The consequences of a loss or betrayal of trust show just how crucial this network is. When trust proves to have been misplaced, the result can be a terrible sense of betrayal, of bitterness.

A British mother whom I shall call "Mrs. Owen" explains the terrible frustrations of life without adequate communication and goes on to describe her loss of faith in the professionals she had trusted:

> But she [her daughter] had no ability to question which was causing the most appalling frustration, I mean it was, what I put down as electrical storms. She would just get to a point where so much had gone in and she would just explode with temper. I mean the behavior has to be seen to be believed. Unless somebody has seen the kind of behavior this invokes in a deaf child, they just wouldn't believe it. She would have these whirlwind days, where she would just literally go through the house just throwing things, sweeping things off the side, and breaking things, hitting people. And this was just because she didn't have any access to communication, because all the time, all the professionals we had been in contact with, their whole policy and their whole sort of ethic was oralism. You know, and because you don't know any better, you think that they know what they're talking about, you go along with it. And it wasn't working. it just wasn't working.[14]

Ultimately the Owens decided to look for a school that accepted the use of sign language and to learn sign language themselves. The transformation was dramatic:

> MRS. OWEN: Within eight weeks she had picked up so much sign language. All the tempers had gone, she had calmed down, she . . . it was the most dramatic change that you can imagine. You know it was like suddenly gaining a child or even finding a child. . . . I just didn't enjoy her as a child at all because it was such dreadfully hard work. But all of a sudden within eight weeks it was like gaining a daughter that you could communicate with. And by then I had started to fairly intensively learn sign language. . . . So all of a sudden the problem wasn't really a problem. It was so much easier. Well, if this can happen in eight weeks, why didn't we use it from the beginning? Why didn't we use it from the beginning?
>
> INTERVIEWER: How does that make you feel?
>
> MRS. OWEN: I felt incredibly bitter about it. I felt very, very bitter about it, because I thought if only somebody had had the foresight or the . . . be broadminded enough to say yes, I think you should try sign language from an early age.

Trust in the medical profession, though less unquestioned than it once was, is still a characteristic of modern society. So too is the overwhelmingly hopeful and expectant attitude toward medical advance that dominates media reporting of new medical technology. It is because of these predispositions that the ethical doctrine of "autonomous choice" is so powerful a discursive resource. It provides proponents of new interventions, including cochlear implantation, with a means of deflecting critique from advocacy groups. In the absence of pressure from such groups, most patients (or parents) will come to share their positive view of the benefits of a new technology.

"A Little Hearing . . . A Little Hope"

In 1996, Lucy Yardley, a psychologist, arranged interviews with a sample of parents in England who had considered cochlear implants for their deaf children. Mrs. Owen was one of those parents. The interviews were conducted by two students under Lucy's supervision. The timing is important. By the mid-1990s, a specialist pediatric implantation center had been established at the University of Nottingham, and pediatric implantation was entering a period of rapid growth. Although the British Deaf Association still opposed pediatric implantation (though its position was far from extreme), the position of the National Deaf Children's Society was now one of cautious acceptance. Public controversy was fading. In the interviews, parents talked about what was involved in their deciding for or against the implant: the emotions and the things they had considered.

Listening to these parents talking about their experiences, it is important to focus not only on their hopes and expectations but also on their experience of making up their minds as to what to do. The picture that emerges is very different from that conveyed by the Nottingham study, which was based on interviews with parents of implanted children and conducted four or five years later. What did "freedom of choice" or "autonomy" mean to these parents? A parent might think, "It is my responsibility to look at the options there now are and decide. I have to decide on the basis of my child's present interest. If she's going to blame me later . . . well, that will be hard but I have to live with that possibility." If I read the bioethics literature, I might feel more secure in the knowledge that I was behaving well. But a much more intractable problem then follows. How can a parent know what his or her child's best interests are? There is no shortage of people who seem certain. The cochlear implant centers know and encourage parents to trust them. Having lived with deafness all their lives, deaf adults also know. The neighbor knows because she saw something on television. An abstract ethical principle may help, but bombarded by contradictory advice, it does not help very much. The decision is the parents', but only in the formal sense that they bear ultimate responsibility.

The interviews make clear that this was not only a complex issue but a highly emotional one. Feelings of guilt, hope, and a sense of responsibility: all were deeply involved. Many of these parents were aware that feelings of guilt may have been driving them to offer the child too much by way of compensation. Two mothers expressed these feelings five years after their children were diagnosed as deaf. Here is Mrs. Smith: "You've got to spoil a deaf child or anyone who's got something wrong with them, for the guilt factor, because there's definitely that thing that you feel guilty, that you've actually brought them into the world. And even though, we've had all of our genetic counseling done and we were told it's just one of those things, . . . but you still feel very guilty." Mrs. Carter put it like this: "There's all this thing about parental guilt because there's no where else to go with it, is there? I mean P is deaf because it was just one of those things—it's us, it must have come from us, it must be a genetic thing. And so there is a temptation to want to make it right, to make it up to him, to compensate something . . . but I mean you can't. I mean a cochlear implant is not a cure, he's always going to be deaf, even if he had a cochlear implant."

Even when feelings of guilt have been dealt with, they can reemerge under the stimulus of outside events. Reactions to the child, for example, can easily undermine fragile coping strategies. Here is Mrs. Smith again:

> I mean I've been in tears before when I've been down the park and I heard this boy particularly saying to his mate—He's going "um, um, um," being friendly, smiling, always friendly and likes the older kids especially boys—and he called his mate over while I was standing there, and said, "come over here, come and listen to her, isn't it funny? Doesn't she sound funny?" . . . I went away from there and I felt absolutely awful. H was smiling, running around—she doesn't know what they're saying about her. I do. I'd rather she heard if someone called her a deaf whatsit. Do you know what I mean? I'd like her to hear that. But the fact that she can't blimming hear it, that's what makes you feel very guilty, very guilty.

The conversations with parents do not show rational "decision makers" weighing up the evidence and "autonomously" reaching a decision. They show people beset by uncertainty, by doubt, by the burden (not the right) of choice. It is while struggling with emotions like these that the parents came to hear about the cochlear implant.

> INTERVIEWER: How did you first hear about cochlear implants?
>
> MRS. LATTIMER: My Teacher of the Deaf told me. . . . he was oral and he wanted B. to try the oral route, but when he couldn't, he ended covering his face whenever he came and so he decided to go over to signing.

So . . . he told me about . . . I think it was first suggested that we go for a cochlear implant by [the teacher].

INTERVIEWER: So the Teacher of the Deaf was the first introduction to cochlear implants? So what did they say about it? What was your impression they gave you?

MRS. LATTIMER: That it wasn't by any means perfect, but it was at least something, that it was something to hold to. You know to give him a bit of hearing, to give him a bit of hope. It sounded reasonable. It sounded as if there was something good there. Something to hope for. He spends his lifetime in silence, so you know, so he spends his life just signing.

The interviews show people whose decisions, and the ways they talk about their decisions afterwards, are the product of complex social interactions. The advice and approval or disapproval of family members, professional advisors, and friends as well as a whole range of other experiences have all played a part. However extensively parents tried to inform themselves, however diligently they sought out and studied relevant documentation, many of those close to them did not. Even knowing that other family members hardly understand what is involved, it may still be difficult to ignore their opinions and their feelings. Mrs. Lewis recalled, "My dad said to me one day—I'll never forget—'Oh look, the local college are doing two courses. There's one on computers and one on signing.' So I said, 'Oh good, you've signed up then?' And he said, 'Yes, I'm going on the computer course.' And that sort of summed everything up. They were too frightened."

Children's grandparents figured in many of these narratives. Grandparents are shocked and saddened by the deafness of a child. Unwilling or unable to understand the complexity of the matter, they were all too happy to absorb the oversimplified message conveyed by the media.

INTERVIEWER: What were your family and friends' reactions?

MRS. HARDING: [Husband's] parents were for it. I think my mother-in-law particularly. She never really got over the fact that her eldest granddaughter was deaf. She was the sort of person that got very flustered and upset whenever you went into any depth, so you tended to not want to say too much. [Husband's] dad would have forked out any sort of money just so that she would be able to hear.

INTERVIEWER: How about friends and family?

MRS. SMITH: Family, like all families I suppose, they haven't got as much information naturally, and their immediate reaction was you've got to have it done. . . . And everybody you speak to, friends, it used to get

> us down because people used to say, Has she had it yet? Is she having it next week? . . . I think about 90 percent of people, including family and friends, thought it was the miracle cure, but we've kept an open mind on it from the word go.

However much parents tried not to let themselves be guided by the simple picture painted by the mass media, refracted through the views of others that picture still influenced what they did.

> MRS. DONOVAN: Why did I think it would give him his hearing back? Well, because it's like a, it's made out to be like a miracle cure, isn't it?
> INTERVIEWER: Is that from the media?
> MRS. DONOVAN: Yes, from the T.V., if you see it on telly that is exactly what it is, really.
> INTERVIEWER: Do they just show the good sides of it?
> MRS. DONOVAN: Yes, they didn't show any bad sides of it.

Mrs. Carter again: "One or two friends have said, you know, 'Has he not been able to have a cochlear implant?' They've worded it in such a way as we've been denied something. So I think the press do . . . it is portrayed as 'this is a cure for deafness.'"

The ethicist's dilemma—that of balancing the right to free choice against the best interests of the child—seems to have been no dilemma for parents in practice. There was a fundamental dilemma, an uncertainty, but it was a different one. How do I know what is in fact in my child's best interests? Should I as a parent try to weigh up the evidence and decide what would be best for my child at this moment in time? Or should I try to put myself in my child's place, years from now, and try to understand what that future he or she would have wanted me to do? Or, put differently, "Should we act in their best interests, or as they would have directed us to act?"[15] That could be different. It depends on how we think our child is likely to grow up. Some of the parents were very well aware of this dilemma. A few of them tried to deal with it by involving the child directly in the decision. Mrs. Harding was the mother of a ten-year-old girl:

> At the time I was in two minds. I mean I felt that at the end of the day it had to be S's decision, because it wasn't me that was going under the knife and having to put up with the electrodes and things like that. But if that had been what she wanted, then I would've been happy to go along with it. My husband, again, was I think in two minds, but he was more for it than against. So you know, we were in two minds but I think he was more for it than against. But I was very undecided, I think. I think if she had been the same then it would've taken us a long time to

> make a proper decision. In a way I'm glad that she was the one who had to make the final decision, because I don't think I could've done.

Most of the children were younger than this and perhaps would not have been so able to make their own preferences clear.[16] What is the child, at age thirteen, eighteen, twenty-five, likely to have wished? How do parents build that assessment into the process of deciding when the child is still young? Some parents did try to consider what their child might later prefer. As one put it, "I didn't want her turning round to me and saying, well, you took this decision when I was very young and I don't want it." Here is Mrs. Owen, whose child was eight at the time:

> At the moment, she is solely my responsibility and really how I feel . . . and if I feel something is right then at the moment that's how it has to be, because no child of eight—even a normally hearing child—has the balanced ability to take decisions for themselves, whereas by the time she's sixteen, seventeen of course she will have. And of course she will know more about the deaf world, about the world in general, and her own perception of how she wants to accept it, or be accepted, or feel that she has to cope with and know what kind of prejudices and kinds of stresses there are and if she felt that that would help her better, then . . . I mean if it had improved, then yes, I wouldn't be against it, but if it was still in exactly the same stage I think I would feel very unhappy about her making the decision to have it done, but I still wouldn't stop her if she wanted it, not at that age.

The Smiths looked at the same issue differently:

> MR. SMITH: I class us as responsible parents and I think this decision has to be made at an early age. I mean somebody in a similar situation to us with a daughter the same age, they're terribly against it . . . they don't want to put her through the operation and they think that the child will make her own decision about thirteen . . .
>
> MRS. SMITH: No, about eleven or so . . . But the majority of children at that age . . . I mean, she's gone all that time in that situation, she's not going to say, "Oh yeah, I want the five-hour operation and my head shaved off and a great scar, yeah," is she? I mean, in a way, the ignorance is bliss . . . this is why it's so difficult, we have to make the decision because she can't make it, and we hope that she's going to say, "Thank you mum and dad, for putting me through that, even though it was quite a trauma."

The Smiths recognized that left to grow up in the Deaf community, their child may well come to reject the idea of a cochlear implant. But while Mrs.

Owen respects the insights and experiences that might lead her child to grow up to reject implantation, the Smiths see those same insights and experiences in another light. Growing up in the world of the deaf, the teenager cannot be trusted to make a sensible decision. Moreover, inaction, delaying implantation, can itself be viewed as depriving the child of choice at a later date, as Mrs. Freeman explained: "I didn't want to be in a position when he is older for him to say 'Why didn't you let me have it?' At least this way he's had the implant, hopefully he will learn to speak and to use it well, but at the end of the day he can choose not to wear it if he doesn't want to. If he decides he wants to stay in the deaf community and he doesn't want to join the hearing community, then he's got that choice, whereas, the other way round, at a later stage, if he said to me why didn't you let me have it, it would be very hard to say I didn't want you to have it, or something like that." But while this argument makes sense in the context of a child who (like this mother's child) is learning sign alongside speech, it is noticeable that the same urgency to ensure that the child has access to sign language is seldom expressed despite the social, educational, and emotional deprivation that can result when a deaf child cannot use this mode of communication.[17]

The difficulties these British parents faced in deciding what was in their child's interests, how (if at all) to involve the child in decision making, and coping with competing advice were not unique to Britain. The interviews conducted with American parents by Gallaudet University sociologists John Christiansen and Irene Leigh a few years later show similar reactions. In their study too, parents suffered from the conflicting advice they sometimes received (fifteen of their twenty-six parents referred to this).[18] The American parents, like the British ones, differed in how they approached the issue of involving the child in decision making.

> Well, it's my child, and it's our job to do the best we can for our child, so I believe that until they are mature, that we have to do the best that we can for them, just like you do with other things. Parents don't let their children do other irresponsible things, either. (American father of six- and four-year-old children each implanted at age two)[19]

> We did not want to make the decision. We wanted her [the eleven-year-old] to make this decision because she's old enough. (American father of a twelve-year-old girl implanted at age nine)[20]

Imagination Work—Understanding the Deaf Perspective

Many parents of deaf children in this study and in Christiansen and Leigh's were aware that there were different views regarding pediatric implantation.

But consider how, for many families, it worked. The doctors and other professionals on whom they depend tell them about the cochlear implant: from them they get the point of view "There is no certainty that it will make him hearing, but it will help. It will make him hard-of-hearing. We know because thousands of children worldwide have been implanted and we know how much they've profited. It will help him enormously in acquiring (spoken) language, although we can't predict exactly how much it will help him. To get the most out of the implant it is better to avoid using sign language (though that is of course a perfectly respectable means of communication for those who cannot speak or hear)." The parents contact the cochlear implant center, and are provided with all kinds of information regarding the implant. The selection procedure is explained, the operation is explained, and the long and painstaking process of (re)habilitation is explained. The brochure tells them about other children who have profited from it. They are put in contact with another family with an implanted child living not far away. Talking to other parents who have coped with the same problems and uncertainties is a great relief. The parents can share their anxieties and their uncertainties with people who know what it is. Their child is clearly doing well. The information provided also explains that deaf people do not see themselves as in need of repair. They manage perfectly well with the help of sign language, and they are able to lead fulfilling lives. The parents are told that, from the Deaf point of view, sign language is an acceptable and alternative means of communication to which their child has ready access. Many of the parents feel they now know enough. Their child is put into the application process. The professionals they have met clearly have the children's interests at heart, and trusting them comes naturally. The family with the implanted child that they met—a similar family—has been through it all and has only positive things to say. The things that the Deaf community are saying begin to sound judgmental and uninformed, even threatening. Trying to do their best for their child, they were being accused of abusing it! Some of the American parents whom Christiansen and Leigh interviewed referred with considerable resentment to what the National Association of the Deaf had had to say.[21]

As a parent, I know I should be properly informed before making a decision that will have major implications for my child, for my family, and for myself. Ethical-legal doctrine tends to stress precisely this: consent should be informed. But what do I have to do before I can feel that my decision was in fact "informed"? Is the kind of awareness of the options that the implant center's brochure contained sufficient? I think not. After all, talking to a similar family with an implanted child makes me realize that what I am really looking for is help in imagining what my deaf child's life—and our family's life—might be like in years to come.

Despite the emphasis on providing comprehensive information, the problem is not one of information alone. It is far more fundamentally a matter of imagination, of trust, and of the social nexus within which relationships of trust are forged. The saliency of information depends upon trust in its source, and its interpretation demands imagination work. Comprehensively weighing the alternatives should involve trying to understand what the Deaf community has to say regarding the lives of deaf people. In practice, only a few parents in these studies wanted to know more about the lives of deaf people or the Deaf community's views of cochlear implantation.[22] Why was this?

> MRS. SMITH: I think the reason why we haven't been to any Deaf clubs like that is because we don't want to treat her any different from a normal child. Because she goes to school, she goes to a normal school and she's taken out of that class and she's in a special support unit with other deaf children. So she knows she's slightly different from the rest of the class, which is fair enough, she is. But I don't want her to actually push her to a degree, force it down her throat she's deaf and she's just got to mix with deaf people. We want her to mix with all walks, I think that's the answer to that one . . .
>
> MR. SMITH: Going back to a few years ago, before all this, that was the way. Whenever did you meet or hear about a deaf person? And you wouldn't even dream of trying to converse with them, because there again, most of them just grunt and groan at you, or sign away at you and not open their mouths. And it's quite sort of off-putting.

Various rationalizations of this unwillingness to contemplate the Deaf alternative were given: denial, a fear of Otherness, of abnormality. One of the English parents, Mr. Bottomley, put it like this:

> Myself and [wife], neither of us have been very enthusiastic about sending P along to these meetings [for deaf children]. And I think it's because we don't want him to identify himself as a deaf child. We want him to have this sort of consciousness that he is a child, a normal child, who happens to be deaf. And so if we group him together with other deaf children—I know it sounds as though they've got the plague or something, which obviously we know they haven't—we just thought it might sort of reinforce in him that this is a very, that he's a very definite type of child, do you know what I mean? We thought it might stereotype him in a way . . . it may be partly because neither of us wants to subconsciously accept that he's deaf, I don't know.

Some felt they knew what deaf people thought without talking to them. Christiansen and Leigh quote one father who said, "One of the things we noticed with that *60 Minutes* program, with Caitlin's story, was that when Roz [Roslyn Rosen][23] was talking, she wasn't able to communicate unless she had an interpreter, whereas little Caitlin was doing everything on her own, and it seemed so black and white, like, What's the matter with this woman? . . . Why would you want to try and put a roadblock up for somebody that wanted to go out there and do this? And it just boggled our minds."

Cultural stereotypes, a fear of the Other, played a role, as did the way sign language and the Deaf community were presented by people parents trusted, or on whom they were dependent:

> INTERVIEWER: Who were the people then who were saying that signing is bad?
>
> MRS. LEWIS: Well . . .
>
> INTERVIEWER: Obviously don't say any names.
>
> MRS. LEWIS: Oh gosh, no. Well, for an example, I went to the school—the local school that my elder son goes to—just to see how they felt about things, how C. would cope in their school, what they would offer him. And the parting words from the headmistress were "Remember Mrs. L, it's not a signing world out there." And that just sort of went through my head. And for some reason it just put me off the school totally. Because I thought that's so ignorant. And so shallow.

The implant teams, like all professionals, cultivate patients' trust. When Alice visited the implant team shortly after the conversation I described earlier, her anxieties were brushed aside. The implant teams wanted her to have confidence that they had her child's interests at heart. And the point is, of course, that when we put ourselves in the doctor's hands, we need to and we want to trust him or her. Trust seems justified, and not by previous experience alone. Everything the medical team does and says is supported by the authoritative discourse of science, by the symbols of medical and scientific success, and by what we read in the newspapers and see on the television. The Deaf compete for parents' trust on an uneven playing field, just as they do in their struggle for political acknowledgement of their point of view.

Imagine that a few parents do overcome their fears and go along to the local Deaf club. Nobody speaks much, they cannot follow the signing, and it is not easy to make contact. They learn that it is possible to enroll for a course in sign language and a few do. Perhaps many do. As they get on with their sign language lessons, they get to know a few deaf people who are pleased to see them there. Gradually, they feel they are coming to understand what the life of a deaf person is like and what lies behind the objections of the deaf to pediatric implantation. They realize how little most people (the doctors they had spoken to, the teachers of the deaf, the friends and neighbors) know about deafness or the world of the deaf. Deaf people are scornful of doctors' ignorance of deafness. But the more the parents come to understand, the more difficult it is to know what to do. In both the United States and the United Kingdom, there were some parents who had made an effort to approach the local deaf association but had felt inhibited (by their own inability to sign) or even rebuffed. There were yet others who had gone further: learning sign language, developing contacts with the Deaf community, but who despite knowing deaf objections to implantation were nevertheless considering, or had opted for, cochlear implantation. Here there is no question of rejecting deafness or of being put off by difficulties of making contact. As the environment becomes more complex, as the value of sign language becomes apparent, the choice becomes all the more difficult.

Mrs. Lewis experienced a transformation in family life like that described by Mrs. Owen but was still undecided about implantation. What Mrs. Lewis had to say points to the enormous gap between the simple abstraction underlying what they had been told and their own experience:

> We have for four and a half years—and this is no disrespect to any parties that have been involved—but it's always been a case of, he has got some residual hearing and he will learn to talk, you just need to keep plowing the information in and it's going to come back. So we had spent four and a half years talking, showing, doing, and the frustrations, the older he gets, were horrendous. Simple things like drink. You know, he'd come in and you'd have to get every cup out of the cupboard to find out which one he wanted, go through every drink. And then we might not have what he wants, but he can't explain what he wants. So, we have another tantrum with throwing things and screaming. And then it suddenly dawned on me, when he was one and a half years old—he needs something more. And if that channel is signing, then that channel has to be signing. Whilst I've been opposed to it thus far, only through, again, the information that has been put my way: "it's not a signing world out there and blah-de-blah." I suddenly realized that that's been the answer. And it's made our lives so much easier. . . . And I feel angry, for want of a better word, that I haven't done it so much earlier. And I feel that I've wasted a lot of time. And I feel that I've wasted . . . I feel guilty—because C. deserved something before now. So, yes, we are learning to sign.

The Thomsons had decided for an implant, in the hope that their child would have access to two worlds. They were trying to imagine a future in which their child would be able to move between these two worlds: "It's more in K's interest. Ours as well, so that we can understand deafness. It's all well and good having a deaf child, but you've got to understand the way deaf people think and feel. You can't just sort of ignore it, because it's always going to be there. And the more we know about it, the better. I mean these are the people who have had the experience all their lives and they could tell you the problems you're going to come against. The pitfalls, how they feel, if they get depressed or whatever."

The narratives presented above show parents—some parents—trying to imagine possible futures for their deaf children. Fears, stereotypes, and difficulties of access stand in the way. There is another aspect of this "imagination work" to which anthropologists Rayna Rapp and Faye Ginsburg, writing about disability, have drawn attention.[24] Their illustration of it starts with a television program *What Are You Staring At?* aired repeatedly on an American children's cable network in 1999. In the program, children with a range of impairments—hearing and visual, cerebral palsy, burn injuries, Down syndrome—talk about their lives with two "celebrity crips," one of whom was the late Christopher Reeve, Superman, paralyzed in a riding accident in 1995. Then we are introduced to ten-year-old Samantha Myers, a New York girl suffering from a rare condition known as familial dysautonomia (FD) that affects all forms of bodily regulation. Deeply affected by seeing the program, Samantha told her mother that she wanted to go on the program and talk about *her* condition and *her* life. Within a few months, Samantha was working on a five-minute segment for the show, eventually broadcast in April 2000. "Of greatest significance to Sam was that so many FD kids were able to see another child like them on television. She was deluged with e-mail from families with FD children around the country who were thrilled to see an image and story that for once included their experience."

Out of Samantha Myers's experience—her sense of kinship with the disabled children she had seen on television and her desire to join with them—Rapp and Ginsburg extract their notion of "mediated kinship." With this term, they want to help us free ourselves from a deeply ingrained image of the "normal family" and "normal family life." Normal family life does not have to refer to two kids, one boy and one girl, a couple of years apart in age, coming home from school to Mom's newly baked apple pie. All sorts of families can be "normal" and loving, and these include families in which one member is in some way different. Most families still have to understand that this is so. Kinship does not have to be about blood and genes and facial resemblances. Families must learn, just like Samantha's family learned from the way she related to all those other children across the country. That of course is exactly the transformative potential of the mass media to which Rapp and Ginsburg draw attention: "It is not only the acceptance of difference within families, but also the embrace of relatedness that such models of inclusion present to the body politic that makes these spaces potentially radical in their implications. As sites of information and free play of imagination, these cultural forms help to create a new social landscape."

The notion of a "Deaf community" may be strange and even threatening to a family trying to come to terms with a newly diagnosed deaf child.[25] That adult deaf people might have useful tips is one thing, but the notion of a strange "community"—distant, different, difficult to approach, claiming some sort of affinity with my child—is something else entirely. Such a notion is threatening, disruptive, precisely because it seems to challenge established ideas of the integrity of the family. The family—kinship—in Western industrial society at least is defined by birth and by blood. How can such feelings be reconciled with the claim that my deaf child has—needs—affective meaningful kinship ties with people I have never met and whom I find distressingly different? It is these established ideas of kinship that make it difficult and that Rapp and Ginsburg want us to question.

This analysis suggests that, in Britain in the mid-1990s, only a minority of parents tried, or were able, to make "thoughtful and informed" decisions in the sense of truly imagining the available alternatives. This is unlikely to have been a uniquely British phenomenon. Just as in regard to the political decision making discussed in earlier chapters, the "positive" and "negative" perspectives on the implant reached them in very different ways. Detailed information on the implant and what it might offer their child is provided by the implant teams: enshrined in a web of institutional practices and carefully cultivated relationships of trust. These parents are accustomed to trust the medical and other professionals on whom they are so dependent. There is reason to believe they have children's best interests at heart. In the cases where professionals appear to have betrayed parents' trust, the sense of betrayal is powerful and painful. By contrast, what deaf people have to say, for those who try to find out, bears the stigma of its origins at the margins of society. Confronting deafness for the first time, few parents are likely to feel they can or should trust these people claiming affinity with their child; nor do they really believe in the notion of a Deaf community.

There is also a difference in cultural dispositions that shapes parents' receptivity to the conflicting messages. The mass media, emphasizing the promise of medical progress, reinforce the hope of restoring normality: a normal (that is, hearing) child, a normal family. The personal stories carried by the media are highly persuasive, particularly in the case of celebrities whose fame or glamour encourages readers or viewers to identify with them. Heather Whitestone, who in 1994 was the first deaf woman to be chosen as Miss America, is one such example. When Whitestone was fitted with a cochlear implant, her picture and story came to figure largely on the manufacturer's website.[26] There was a link to *USA Today* too, where she appeared under the headline "With Implant Miss America May Hear Again."[27] Like the headlines that the early implant operations attracted in the United States, Britain, and France, words like this provide the kind of reassurance that people expect of medicine. Media representations rarely encourage an imaginative contemplation of alternative ways of "being in the world."

How could it be different? Rapp and Ginsburg suggest that what is needed is something they call "rewriting kinship." They draw a distinction between the public discourse surrounding disability on the one hand and "the daily and intimate practices of embracing or rejecting kinship with disabled fetuses, newborns, and young children" on the other. The first is a matter of ethical debates about reproductive choice, of legislating for access to buildings, to schools, to employment. In other words, it is comparable with the political claims of the Deaf community, with which many parents had great difficulty. The second, rooted in changing daily practices in the family, in the work of clarifying and articulating what a disabled child might mean for the family as a whole, is more fundamental. At the heart of the notion of citizenship championed by the disability rights movement is the integration of disability into everyday life. The shift from exclusion to inclusion, reshaping the possibilities of life with a disability (or as a deaf person) has to have its roots in family life. Rapp and Ginsburg argue that "public storytelling," testifying to problems overcome, lives successfully led, is crucial to bringing this integration about. To be sure, some parents are likely to be better able to do this imaginative work, conjuring up for themselves a picture of how life with a disabled (or deaf) child might be. Differences in material circumstances also play their part. Some families are in a better position than others to accommodate a variety of special needs and new caretaking tasks. Working-class women, in this account, realistically doubt that the social support they would need to cope with a chronically sick child or a child with a disability would in practice be available. Their material relief, their new possibilities, will be created by the cultural work, the public interventions, of their better-off peers. Rapp and Ginsburg provide a number of telling illustrations of what they mean. One concerns Emily Kingsley, a scriptwriter for the children's television program *Sesame Street*, whose son Jason was diagnosed with Down syndrome. Instead of putting Jason in an institution and "trying again," as her doctors advised, Emily Kingsley wrote him into the script of *Sesame Street*, where he continued to appear for a number of years. It is the work of families like this—"families well positioned for activism"—that, by changing the cultural landscape, makes it easier for other families to contemplate life with a disabled child. It is not difficult to see the similarities to what we are discussing here. Imagining what it would be like to have a deaf child, a child for whom sign language is a first language, is not something that takes place in a cultural vacuum. It depends on the presence of deaf people in the public culture—the soap operas and talk shows, movies, the newspapers, fiction, the Internet—of the region or country in which a family lives and grows.

In trying to understand why it is so difficult for the views of the Deaf community to receive an equal hearing, the justifications provided by the dominant bioethical doctrine and its emphasis on informed consent once more play a role. Bioethical principles obscure what is really at issue.[28] The notion of "information" obscures the work of imagination that putting abstract information to use entails. It is because linking abstract statistics to an individual child's life is so difficult that learning from the experience of other families is so valued. And the principle of autonomy, among other things, diverts attention from the many ways in which decisions are in fact contextual, dependent on networks of trust and dependency and on simplified media representations refracted in the views of grandparents, friends, and neighbors. Other things being equal, the "free choice" of most consumers will be in line with the message conveyed by the media and reinforced by the family members, friends, and professional advisers. It is this context that makes the notion of autonomous choice, enshrined in ethical doctrine, so valuable a resource for those advocating the merits of new medical technologies.

"A Vague Kind of Way to Exist"?

Few studies of parental decision making or of the consequences of cochlear implantation have anything to say about identity. The Japanese study discussed earlier is exceptional in this regard. A number of Japanese parents were concerned by the uncertain identity of their children after implantation. One had been troubled by what she had heard from another mother, herself deaf: a deaf mother "said that there is a hearing world, but children with implants are not necessarily hearing. She referred to those who use sign language as Deaf people, distinguishing the two worlds completely. I told her about our plan for our daughter to receive an implant and she called my child a 'social outcast'" (*aburemono*—a discriminatory term for those who are outside social norms).[29]

These authors note that some respondents "were puzzled by this separation between hearing and Deaf communities, and wondered whether children with implants would form a Deaf identity, hearing identity, or a separate new identity." Studies conducted in western countries rarely refer to puzzlement or doubts of this kind. Most focus exclusively on functioning and performance. Largely this is limited to auditory performance. But even when researchers did try to explore parents' perceptions of the wider consequences of implantation, including in the educational and psychosocial realms, they tended to focus on functioning. For example, in the British study quoted earlier, parents referred to children's improved ability to communicate with and relate to other people and to improved school performance.[30] Questions of identity and belonging that so concerned the Japanese parents (and the Japanese researchers) were not raised.

By contrast, when adult implantees discuss their decision to seek an implant, and what it has meant for them, a more complex picture of the distinction between Deaf and hearing worlds, and their place in or between these worlds, emerges. Beverly Biderman is a university-educated Canadian woman with a good job in the computer industry who suffered from a progressive hearing loss. From the age of ten she was obliged to use a hearing aid but always tried to conceal her hearing problem and never identified with the Deaf community. At the age of forty-six, after being deaf for more than thirty years, she decided to obtain a cochlear implant. She had looked into it earlier and in the 1980s had been tested at the House Ear Institute, where the single-channel system was then being fitted. She had not been a suitable candidate. A decade later she was trying again. In the book she wrote about her experiences, subtitled "A Journey into Hearing," Biderman explains that she told few people of her decision to go ahead with the surgery. "I was afraid of appearing pathetic, of seeming to be chasing cures. 'What?' I could imagine people asking. 'After all these years of deafness, are you still looking for a miracle? Haven't you accepted it yet?' I had invested so much time and emotional energy into adapting to my deafness since childhood that it was hard to acknowledge that yes, it was still a problem, and that I accepted it only with great reluctance."[31]

What she calls her "journey into hearing" began in July 1993, with the switching on of her processor six weeks after surgery. She recounts her "journey" not so much as the story of an auditory triumph but more as the painful remaking of an autobiography. She had to confront the anger and the frustration her deafness had caused her, especially as a teenager desperate to enjoy the music her peers were enjoying and desperate to conform. "I began to understand the magnitude of my efforts to pass as hearing, when I got more hearing and began to see how little I really had before. I was then able to forgive myself for my deafness." This remaking of autobiography brings grief with it. Grieving for the teenage girl she had been was "a belated but necessary step for me finally to accept my deafness."

Trying to learn to hear involved much more than just a training program. What had previously just been there, a frustration and an inconvenience but backgrounded, was now brought sharply into focus. "Am I becoming a hearing person?" The effort of interrogating her status as a deaf person is wrenching: she describes it as an "unraveling" of the fabric of her life. Whatever her prior dreams and expectations, Beverly Biderman has not after all become a hearing person. She is a deaf person, she explains repeatedly, but one who can communicate effectively via spoken language, thanks to the implant. She is confident that her deafness will become less and less of an inconvenience, thanks to time, experience, and further technological development. Her autobiography, it seems, has not been wholly rewritten. Her knowledge of what it is to grow up deaf leaves her with considerable sympathy for the Deaf community's opposition to cochlear implants. "I understand this opposition with regard to the Deaf asserting that their lives are not so terrible as hearing people looking at their world from the outside may think. Strangely, I even find myself, almost against my will, extending my sympathy and understanding to their opposition to cochlear implants. I find myself comprehending it on a level where my own pain about deafness resides."[32]

In Britain, the National Cochlear Implant Users Association published a small book containing the personal testimonies of fifteen cochlear implant users, of all ages and with very different experiences of deafness.[33] Lisa Thomson, for example, started to lose her hearing at the age of nine, and became profoundly deaf, overnight, at age twenty.[34] "At the time," she writes, "I remember thinking I couldn't have cared less if I never heard another person speak as long as I could hear music." Thomson goes on:

> A few months after becoming profoundly deaf my speech therapist suggested a cochlear implant. Like hearing aids I refused. At the time, I thought that by having it I was giving up my chance of future technology finding a cure. The idea was to change by the following March. Heather [the speech therapist] had got me a place at a job club for the deaf in London and I began to mix with other deaf people socially. At first I was on top of the world and could hardly wait for the fortnightly meetings of almost 200 deaf people at a London pub. Despite this I was very aware of the differences between the two "worlds." Those born Deaf were proud to be so, I wasn't. I made many friends and even had a deaf boyfriend but it wasn't enough.
>
> I had changed my mind about the implant and it quickly became apparent that my opinions weren't shared. What made it so difficult was

> the fact that the only positive opinions I received came from the medical world. I got into many arguments, including with my boyfriend, and I realized in order to make any real decision for myself I had to come out of the deaf world. It was one of the hardest decisions I ever had to make and I spent most of the time in tears.

A year later, she explains, "I have plucked up enough courage to go back to college and from there went on to university to do a degree in psychology." With great courage and with difficulty, Thomson seems to have picked up the threads of her life. There is no doubting her delight in her implant, but it is worth reflecting for a moment on the things that, for her, marked the long disruption of her life. The loss of music, not of speech, marked its beginning. Finding the courage to resume her education marked its end. These are not the criteria of audiological assessment. Nor is her story easily accommodated to the familiar "woman hears again thanks to miracle cure" newspaper headline. Thomson concludes by remarking, "I wouldn't go so far as to say I am proud to be deaf but I am not ashamed of it." Although she resumed the life she might have led had she not become profoundly deaf, she remained deaf.

The cochlear implant helped these people hear more, so that they can cope far better with a world full of sound and full of spoken language. But did it make them hearing? From a strictly medical or audiological point of view, it is simply a matter of whether their hearing thresholds were raised sufficiently. Lisa Thomson and Beverley Biderman, both of whom are clearly delighted with their implants, nevertheless resist viewing their experiences in such simplified terms. For both these women, being deaf entails far more than the absence of hearing. Biderman's identification with the world of the deaf does not spring from participation in Deaf culture and the Deaf community. It comes from the pain, the memories of exclusion, frustration, and grief that, she knows, she shares with other deaf people. To be able suddenly to hear is not to escape the whole of her past life. Biderman and Thomson can now function more effectively in the hearing world, but they are still deaf.

For Beverly Biderman and Lisa Thomson, being deaf is more than not being able to hear. For each of them it entails memories of grief and of exclusion, personal histories of relationships with friends and family, ways of knowing the world. Starting from their own lived experience, as the Deaf community does, neither Biderman nor Thomson view the deaf/hearing boundary in simple audiological terms. Biderman explains that, for her at least, it is not possible to cease to be deaf. An improvement in audiological threshold is by far not enough. For her, the boundary between deafness and hearing is a complex region, marked by values, memories, histories, and commitments, and to be crisscrossed in many ways. For a deaf person whose social life is in the Deaf community, who uses sign language at home and with intimate friends but spoken language with colleagues and strangers, this is all the more true. If they discount their experiences, look simply at their audiograms, Biderman and Thomson might say, "Yes, now I'm a hearing person." That is the way people are encouraged to think about it, in medical terms, as a matter of hearing thresholds. Neither of these women is willing to reduce the complexities of her emotional and social life to a simple measurement.

Nicole is a French journalist whom I talked to in Paris in 1993. She had become totally deaf at the age of twenty, more than a decade earlier. Very well informed, Nicole had gotten to know of Chouard's work and of the implant he had developed. That was not for her: she had not been impressed by the device and had not approved of the way he worked. But when a few doctors in France began to work with the Nucleus despite financial inducement to use the French system, Nicole decided to visit one of them. He explained to her that the duration of her deafness was a major disqualification. Deafness, silence, had become part of her, part of her personality. It was impossible to be sure how much benefit she would get from an implant. She would have some benefit: there would certainly be situations in which she would be able to hold conversations, but she must not expect too much. She should think carefully about whether she wanted to go ahead.

Unlike Biderrman, Nicole was much involved with the Deaf community. Making use of her journalistic skills, she had previously set up a Minitel-based information and discussion service for deaf people in France.[35] Nicole decided to publish a diary of her implant-related experiences on the Minitel service, recounting day by day the tests she underwent, what she experienced, what she felt. The test results were good and she decided to go ahead with the implant. This announcement provoked all kinds of reactions. Some accused her of being ashamed of being deaf. But no, that wasn't the case, she explained. She had become deaf, and though she hated deafness she would remain deaf. Most deaf people seemed to understand her feelings and wished her well.

One of the hardest things was the period after the operation and before the implant was switched on. "You've got all the inconveniences, the shaved head, you're exhausted. The anguish of not knowing if it'll work is unbearable . . . Even though Dr F., who'd filmed the operation and had showed me the film, because it had been a perfect operation. He told me that he was sure that it'd gone well. Well, I was still scared, still feeling really low. The psychologist had said to call whenever I wanted. I'd been warned of the risk of two months depression."

Like Biderman, Nicole well understood how media presentation of the implant scares and hurts deaf people. She too sympathized deeply with their fears. She told me of a recent television broadcast about deafness in which she had participated. "There was a row, because Dr. Chouard had sent one of his collaborators, Dr. Fugain, who, in an extremely human debate about tests, a most interesting debate—and there were Deaf people in the audience—just wanted to 'sell' the implant and was saying that everyone should be implanted . . . Two representatives of the Deaf . . . said that they were getting scared: that people were trying to exterminate the Deaf. What right have hearing people to decide that a deaf child is an invalid child? For us, being deaf isn't a handicap. We have our world."

These are familiar arguments. Nicole continued:

> You have to remember that compared to the U.S., where the Deaf have long enjoyed a certain respect, in France it is new. There's a lot of resentment. They have a strong desire for revenge, so they are extreme in their views. Really. But I understand them because they have really suffered. I was asked to talk about the implant and I said what I think. For people who become deaf like I did, it is fantastic. I can hear. It's a miracle. But I am still a deaf person. I am a deaf person who can hear thanks to the implant . . . Dr. Fugain was there and said that everyone should be implanted and so on. I said that no one has the right to say such a thing. OK, I adore my implant. But no one has that right. Perhaps because I understand too well. I imagine my daughter being deaf. I wouldn't accept it. Oliver Sacks spoke after me, and said that I was right. He told about a blind person who'd been given his sight back and committed suicide. . . . confronted with the reality, they terminated the broadcast, because in fact it was Dr. Fugain who was under attack.

Nicole too is a deaf person who can hear but who does not always want to. Not only does she move socially between the worlds of the deaf and of the hearing, but bodily, physically, she also moves between the worlds of sound and of silence. What she had to say sheds a different light on what it can be to be both deaf and hearing.

> I'm deaf in the morning when I get up. It's funny really. Even if I adore everything that I can hear, especially holding a child . . . I get up, make my coffee, talk with my daughter, take her to school, buy my newspaper, take the subway, get to my office, say "good morning" to everyone, sit down, and then switch on my implant. It's funny, but a different rhythm would be too exhausting. When I get back in the evening, I keep it on as long as possible. I have the habit of taking a bath in the evening when my daughter's asleep, so at about 9:30 . . . I take it off and don't usually put it on again. Except when I'm having a discussion with my companion . . . It is me who chooses the moment for hearing or not.

Nicole has had conflicts with doctors on this point. The reaction is typically one of "that's because you were deaf so long. Deafness has become part of your personality." But she sees it as a matter of rights and of human needs. She asserts, "If you have a need to remain in silence, then remain in silence," a view that does not endear Nicole to all who work with the cochlear implant.

If the lived experience of implant users and the stories they tell are taken seriously, a far more complex picture of the effects of implantation emerges than that captured by audiological measurement. The experiences Nicole recounts are different from Beverley Biderman's because she lives in a different world. Paris is not Toronto. The process of getting an implant was different for her. The cochlear implant in France had a unique history, of which Nicole was well aware and which influenced the decisions she made. What a technology does, or means, differs from one society to another. That the benefits of acquiring a high-performance sports car depend on the roads one travels seems obvious. Traveling Italy's superhighways is not the same as jolting along the rutted roads of rural India trying to avoid the cows and cyclists who share the road. It is not only that performance and practical benefit differ from one context to the other. So too does the symbolic meaning of the technology, the message it conveys. This may be particularly true of anything, whether prosthesis or mark, that signifies bodily difference.[36]

But Nicole's story is also different from Beverley Biderman's because of the way she has chosen to live her life as a deaf woman, much involved with the Deaf community. Nicole chooses to use her implant in a way that reflects the place of deafness in her life history and that accommodates her need for silence. Nicole is choosing her own way of moving in and out of the world of sound. It is probably not what most implant teams would advise or expect, but it corresponds with recent anthropological studies that bear witness to the "many ways of being deaf" in the modern world.[37]

What light do these stories cast on decision making by parents? All three of these women had been deaf for a number of years but had been born hearing. Nevertheless, their stories show us something on which few western parents of young children, struggling with a diagnosis and with their anxieties, seem to stop and reflect. In their stories, their decisions to get a cochlear implant, surely by any measure thoughtful and considered, are deeply contextualized. They show how the decision was shaped by their personal histories, by people close to them (partners, friends, professional advisors), and by the intersections of their lives with histories of implant technology. At an earlier moment, neither the technology nor the time had been ripe for them (Biderman had been rejected for implantation at the House Ear Institute years before, while Nicole had preferred not to be treated by Chouard). The stories suggest that changes in context, changes in the cultural landscape, lead to differences in expectations and in how even the most considered decisions are made.

There is another message to be extracted from the three women's stories. None of them regards herself as hearing. Perhaps the fact that they continue to regard themselves as deaf is a matter only of elapsed time: of the adaptations they made over all their years of deafness. I do not think it is just that. Slowly but surely a more complex and more nuanced picture of what cochlear implants do is emerging, one that escapes the simplifications both of early media accounts and of audiometric scales. In their study of pediatric implantation, John Christiansen and Irene Leigh asked parents of implanted children whether they still regarded their child as deaf. To their surprise, as they admit, "the vast majority of the parents said that, in their eyes, their child was still deaf after implantation."[38]

In the early 1990s, when Alice and Francis decided that their son Simon should get an implant, deaf people they knew felt angry and betrayed. It took constant effort on their part to make clear that their choice did not entail a rejection of sign language. They were trying to give their son access to both worlds. They also needed courage when faced with professionals who objected to their less-than-complete commitment to oralism: didn't they want their child to gain the maximum benefit from the implant? By the end of the millennium, in the Netherlands as in the United States, this had changed. Positions have softened. In 2000, the National Association of the Deaf in the United States retracted its highly polemical position paper of 1991 and replaced it by another, more nuanced. "It seems to us," wrote Christiansen and Leigh in 2002, "that the walls between those who support pediatric implantation and those who oppose the procedure are, if not crumbling, at least beginning to show some noticeable cracks. Both parties are more willing to pay more attention to the other side of the argument, even if they may not necessarily agree."[39]

It is no longer so much a matter of "either/or." Over the past few years, more and more young deaf children have been implanted, even though (in the Netherlands) many of them also attend bilingual schools, at least to start with. As it gradually becomes the norm to have a young deaf child fitted with a cochlear implant (and, increasingly with two, one in each ear), it is hard not to wonder if we are not moving further and further away from what might be considered thoughtful and informed choice. How else to explain the difference between the difficulties British parents expressed in this study, carried out in 1996, and the fact that a few years later Nottingham researchers Tracey Sach and David Whynes found that for most parents, the decision for the implant was straightforward and unproblematic?

The simplified messages that have been at the root of the cochlear implant controversy—both that of the Deaf community (a choice for implantation is a rejection of deafness) and that of the medical-industrial complex (implantation offers hearing and normality)—prove to be oversimplifications. What the technology has to offer to the deaf child, or what a medical technology in general has to offer, cannot be assessed, measured, or predicted with the skills and the instruments of medicine alone. What is then at issue, and what needs to be questioned, is the authority of medical expertise and the scope of medical jurisdiction. The medical profession has claimed, and over time has achieved, unique competence in advising the individual patient, or parent, and (in collaboration with industry and expectant consumers) has largely shaped the history of medical technologies like this one. Medical ethics, which has sought to provide both a language and a mechanism for tempering these claims, has proven inadequate. This has been so not only, as Gardiner pointed out, because it has lacked the social power to impose its judgments but because its principlism has hindered understanding of what is at stake.

The conclusion that Tokyo University researchers Okubo, Takahashi, and Kai reach has relevance beyond Japan (and beyond the field of pediatric cochlear implantation). They suggest that parents would benefit from hearing the opinion of deaf adults "including those who insist on Deaf identity and consider the meaning of implantation on the child's identity." They suggest the need for a discussion forum in which the whole range of professions working with deaf children, as well as deaf people, should be involved: "Opportunities to hear about the attitudes of the Deaf community towards cochlear implants, as well as academic research and debate beyond the field of medicine, would give parents considering implantation more accurate information about the actual experiences and implications of undergoing the procedure."[40]

A decade ago, the report of two years' discussion of the disability rights critique of prenatal genetic testing was published. In a remarkably insightful review of what was clearly a tough debate, in which many points of disagreement remained, the authors explain one positive function of the disability rights critique, even for those who could not accept much of it. The arguments, they explain "are intended to make . . . parents pause and think about what they are doing, and to challenge professionals to help parents better examine their decisions. They are intended to help make our decisions thoughtful and informed, not thoughtless and automatic."[41] A decade later, to reflect on Okubo, Takahashi, and Kai's suggestion that parents should be enabled to decide in a thoughtful and informed manner is to be driven to a chastening conclusion. The choice for an implant has become more or less automatic for most parents of deaf children in rich western countries. Despite the rhetoric of "empowerment," developments in practice have led not to more thoughtful consideration but to an easy automatism.

Time passes and children grow up. Parents have to live with the decisions they took years before as surrogate decision makers. They want to feel secure in the knowledge that they did, in fact, act in the best interests of their child. Christiansen and Leigh found that "looking back does not always bring complete peace of mind, even for parents who are pleased with their decision. We often saw expressions of regret for 'what might have been' among parents whose children experience very different degrees of success with the implant."[42] It is odd that ethicists and other scholars devote so much attention to clarifying how we should make decisions while so little attention is paid to how we can subsequently live with them. A decision is made once. We may have to live with a constant, nagging sense of doubt for the rest of our lives. That is why the occasions that cause us to doubt—and in my case, carrying out this study, they were many—are so disconcerting. It is difficult for us even to entertain the idea that we may have failed our child. It is not only a matter of making agonizing choices. It is also a matter of the context in which they are made and the context in which we then have to live with them.

Chapter 7

Politics and Medical Progress

Early development of the cochlear implant was motivated by scientific curiosity, by desire to help people whose loss of hearing caused them suffering, by dreams of vanquishing deafness. Progress then depended crucially on developments extraneous to the field: the emergence of microelectronic components and biocompatible materials, political interest in artificial organs, the increasing familiarity of the idea of an "implant." The implant's early years were marked by a variety of competing designs, each based on a distinctive set of assumptions about its functioning and its intended function. Despite professional skepticism, despite insistence that design of an implant had to be based on thorough scientific understanding, despite disputes regarding the readiness of the technology for use with patients, it was clinical success that attracted attention in the early years from colleagues as much as from the mass media. Cries of "caution" from basic researchers notwithstanding, House's early successes with a simple implant aroused the interest of clinicians in many countries, as well as of potential manufacturers. As interest and commitments grew, networks were extended and FDA approval was secured. New manufacturers entered the field, and they began both to collaborate in establishing a market and to compete for domination of it. Convinced of the economic potential of this innovative technology, governments too played an important role. The Australian government brokered the collaboration between Clark's university research group and the Nucleus medical technology company. Seeing this as a field in which French technology could prove itself, the French government similarly provided support for Chouard. As the technology matured, the initiative passed from the early surgeon-entrepreneurs to industrial corporations, competing for shares of a growing market. Further innovation then took place largely in their laboratories, in collaboration with networks of medical centers using their products.

Neither a theorist of innovation nor an historian of technology would find anything surprising in such an account. It shows the emergence and functioning of a "distributed innovation system" involving hospitals, laboratories, and manufacturers such as economists have found to be characteristic of innovation in the field of medical devices more generally. It shows the triumph of determination, an unwillingness to be thwarted by the fainthearted, ever-growing numbers who have been helped, the evolution of the implant from esoteric device to global product: precisely the history that innovators want to see written and, not infrequently, write. Accounts like this are part of the culture of all medical (or scientific) specialities. They are also part of the common culture: resources to be used politically and socially. Precisely because of the way it resonates with the countless other such stories that form our picture of medical progress, an account like this is in course of becoming the dominant and familiar history of the cochlear implant.

Because of the familiarity of such accounts, because of their hopefulness, they provided the template on which media accounts of the "bionic ear" were almost always based. And these media accounts, in turn, played a vital role in making cochlear implantation possible. Whether in search of research funding in France in the 1970s, of support for local implant programs in 1980s Britain, or of potential patients in the Netherlands or Sweden, pioneers all turned to the local press and television. Encouraging people who had already been implanted successfully to tell their stories in public was a vital part of each of their strategies. The publicity they received, the headlines such as the French "Victory over total deafness" or the Dutch "Deaf woman amazed to hear sounds again after implantation" (that initially drew my attention to the implant) helped surgeons secure the resources and the patients they needed. With their tendency to present developments in biomedicine as "breakthroughs," the mass media arouse the hopes and expectations of people for whom current medical practice can do little or nothing. Stories like these, the sense of history on which they are based, and widespread faith in that history provide medical science with a vital resource. It is not, however, the only history of medical innovation that can be written. Some anthropologists have found their ethnographic work led them to "unsettle" the scientists and clinicians they studied, and that the hierarchical relations between the ways in which illness is known and understood justified their doing so. Something similar can be said of the writing of medical history.[1]

In this final chapter, I propose to sketch out a different, and less comforting, history of the cochlear implant and reflect on its implications.

Evidence and Momentum

For the implant pioneers, in the 1970s, conviction grew out of personal experience. It was not aggregate data, which did not yet exist, that convinced them of the promise of the cochlear implant. It was their work with one or two experimental subjects that convinced them, or the visits they paid to William House in Los Angeles: being able to witness for themselves the results he was getting. But in order to secure the support of their professional colleagues, more than personal conviction was needed. Public reports of the benefits of implantation—sometimes using patient testimonials—were intended to convince largely skeptical professional peers. Pointing to the lack of fundamental understanding of how hearing worked or to the inadequacy of patient testimonials as evidence, many were unwilling to be convinced. Independent assessments based on larger numbers of implanted patients and using quantitative tests gradually became available. The Bilger report of 1977 and the FDA approvals of 1983 and 1984 were vitally important in helping shape a new and favorable professional consensus. Implant pioneers were all ENT surgeons, and it was their initial leadership and their status relative to other professions working with hearing-impaired people that led cochlear implantation to be organized as a surgical practice. This in turn determined what was to count as evidence for its success. With the move to implant deaf children as well as adults, personal conviction once more played an important part. When implantation of deaf children began in France, it provoked considerable criticism and had to be justified. Justification was not, of course, phrased in terms of the vastly greater potential market that pediatric implantation would offer. It was in terms of a "critical period" in language acquisition (though neurophysiologists were less certain such a period existed) and by analogy to the early use of hearing aids. Despite opposition again (but now from parent organizations as well as colleagues), pioneers pushed ahead assuming data would come later. They did, sufficiently so that in 1990 the FDA approved use of the Nucleus implant with children between two and seventeen years of age.

FDA approval gave clinicians in both the United States and abroad confidence that pediatric implantation was safe and effective. For implant surgeons confronting critics, FDA approval was a vital resource, perhaps of greater importance than the evidence on which it was based. Authoritative and expert assessments of pediatric implantation appearing in the early to mid-1990s were sometimes at pains to make clear that implanted children still required special provisions (for example, at school) and would not function as "normal hearing children." Sometimes they pointed out the areas of ignorance that remained: the lack of longitudinal studies, for example, or of studies related to the socioemotional or cognitive development of implanted children. Sometimes, when too much was at stake, they avoided reference to all that was not known. But whether they did or not, neither caveats nor areas of ignorance were allowed to override the conviction that because the implant had been shown to benefit hearing, it merited the confidence of professionals and of parents.

This is crucial. "Improved hearing" was taken as an adequate measure of the benefit of an intervention that, in addition to any surgical risks it entailed, has profound consequences for all areas of a child's development: sense of identity, personal relationships, schooling, linguistic competence, and so on. Linguists, psychologists, and sociologists, not surgeons and audiologists, are expert in these areas. But to have made professional consensus regarding the utility of the implant dependent on the outcome of linguistic and psychosocial studies would have meant subordinating clinical authority to social investigation. The fact that "hearing"—or, to be more precise, speech perception and production—was the measure on which everything depended was a reflection of the medical profession's ability to defend its own jurisdiction. "Hearing" was the variable that could be measured (only) with the tools of medicine and of audiology. In studies of parents' expectations of the implant, mostly conducted post hoc among parents whose children had been implanted, expectations were formulated in similar terms. Parents had hoped that their children would be better able to follow speech and to speak more clearly. In contrast to deafened adults implanted in later life, some of whom viewed the transition between being deaf and being able to hear in much more complex terms, the majority of parents had understood the implant in the terms emphasized by the professionals who had helped and supported them along the way.

Between the mid-1980s and the early 1990s, hearing professionals came to agree that cochlear implantation was of value for deaf people: for adults, for children who had lost their hearing early in life, finally for children who had been born deaf. Evidence gathering did not cease when professional consensus had been achieved. To the contrary, as more clinical centers began work, numbers of publications rose: from an average of 34 medical and scientific publications per annum in the period from 1985 to 1989, to 82 per annum in 1990–1994, 151 per annum from 1995 to 1999, and 217 per annum in 2000–2004. In 2007, 341 publications on cochlear implantation appeared in medical and scientific journals.[2] From the point of view of the individual author or the medical center where the work was carried out, publication enhances reputation and competitive position: for career advancement, research funding, the best qualified staff, and, not least, for patients. For clinicians and institutions at the periphery, far from the centers of medical progress, to offer an innovative new treatment and to describe their experiences in the medical literature is a claim to status and a symbol of achievement.

Publication has an important function for the technology and for the field of practice, but it no longer has to do with the attempt to achieve professional consensus. A vital function of publication, after consensus has been reached, is to "shore up" and protect this consensus and the practice based on it against external challenges. The changing financial climate of health care posed one such challenge. By the mid-1980s, to show that a technology was clinically effective was no longer enough to guarantee a favorable reimbursement decision. As more and more promising treatments clamored for limited resources, decision makers began to look for ways of comparing them. To justify the claim on resources essential to expanding practice, it was becoming necessary to show that a new intervention was not only clinically effective but cost-effective as well. At the first European Symposium on Paediatric Cochlear Implantation in 1992, Professor Mark Haggard gave a keynote address designed to introduce participants to the need for these new kinds of assessments. With any new provision or technology, he explained, of course one must start with clinical research. Safety and efficacy come first. But thereafter something else is required. A political—and in effect economic—case would have to be made for pediatric cochlear implantation, and this would have to be based on a standardized economic calculus. This is where the concept of "quality of life" and then the "quality adjusted life year" (or QALY) came in. Haggard explained to his audience that because quality of life could not be quantified in a satisfactory manner, "We will hence continue to find it difficult to argue from any rigorous quantitative basis that funds should go into pediatric implantation for all children meeting certain criteria, rather than, say, into better support for leukemia patients or a wider range of rehabilitative support for children with mild developmental disabilities." His lecture would "introduce practitioners to some chief concepts of health economics and to the way in which these might usefully inform research and development in the years to come."[3]

Numerous studies along these lines have been carried out and published since 1992. In order to do such an analysis, all kinds of assumptions and simplifications have to be made: regarding whose costs are to be taken into account, how quality of life is to be measured, and (crucial in the case of pediatric implantation) regarding the schooling of children with implants. For example, one study adopts the perspective of the government/service provider.[4] It summarizes all the costs to the public sector of implantation and rehabilitation programs. Costs borne purely by patients (including foregone income and personal expenditures) are excluded from consideration. "Similarly, benefits enjoyed by the deaf community from their own culture and use of sign language are not considered," write the Australian authors of this study. A measure of quality of life is fundamental to the calculation, and in this case the estimate—what was to count as an improvement in quality of life—was made by professionals, not by the people concerned.[5] This study assumes that eighteen months later 65 percent of children would have moved from special to mainstream education, another that 50 percent of children would eventually move.[6] No questions are asked about whether such a move is in the interests of the individual child, or about the implications for the costs and viability of special education for the children who do not receive an implant. In some countries, deaf children are mostly in mainstream education whether they have an implant or not. But the assumption of a shift from expensive special education to cheaper mainstream education is crucial to demonstrating the cost-effectiveness of the implant. As "value for money" became an increasingly important consideration in health care expenditure decisions, studies demonstrating the cost-effectiveness of the technique became crucial to the justification of professional practice. And indeed, almost all of the studies reach the conclusion that, in terms of the improvements in quality of life that the technique offers, cochlear implantation is good value for money.

In order to fulfill their function effectively, publications like these deploy the conventions of scientific writing. One of these conventions is simplification. For example, one recent study, investigating the extent to which Medicare and Medicaid rates of reimbursement represent a financial disincentive to implantation, begins by stating, "Cochlear implants are an established treatment for severe to profound sensorineural hearing loss. Treatment can cost more than $40,000, including approximately $20,000 for a device. Evidence indicates, however, that these costs are typically outweighed by resulting benefits, such as reduced costs of special education and improved quality of life."[7] The article thus starts by assuming that it has already been proven that benefits outweigh costs.

Second, and perhaps more important, is the assumption of generalizability. "Costs are *typically* outweighed by resulting benefits," the article asserts, though it seems self-evident that however fixed the costs (of which the device itself represents 50 percent) benefits will vary, depending, for example, on the costs and availability of services, on the possibility that a person will enter employment as a result of implantation, and so on. The claims made for the effectiveness of the cochlear implant implicitly or explicitly evoke the universalistic character of scientific laws. This is a general phenomenon. If a therapy can be shown to work in one place, then, in the hands of competent doctors, it will work similarly for the designated category of patients in another place. When the FDA approved use of the cochlear implant in deaf children in

1990, it concluded that the device had been shown to improve the hearing of deaf children, with no reference to where those children might have been living or tested. Health policy makers, doctors, and parents are to assume that this is true irrespective of the culture, language, or society in which a child is growing up.

Developing an economic rationale for implantation is not the only issue researchers have begun to address since the early 1990s. There is also growing attention to the gaps that critics had identified: for example, the criticism that too little was known of the effects of implantation on children's school performance or psychosocial development. As consensus was reached in the early 1990s, few such studies had been conducted. Indeed, nearly all the publications on the effects of implantation on children's quality of life or school performance have appeared since 2000, when the technology was already being used on a large scale.[8] To defend professional practice against critique, studies along these lines were initiated and published and could thereafter be used in defending the evidential base of the practice.

In the case of cochlear implantation, these assumptions were significantly challenged only in Sweden. The studies of implanted children that Swedish psychology professor Gunilla Preisler and her colleagues carried out were based on a methodological and theoretical critique of most of what had been done elsewhere. They argued that little of this work was of much value in the Swedish context. The Swedish National Board of Health and Welfare provided financial support for Preisler's work because it agreed that the existing research literature was of little help given the circumstances in Sweden. Almost all the evidence for the effectiveness of the implant had come from countries in which oral education of deaf children dominated. In almost all programs, the cochlear implant had been intended to facilitate deaf children's access to spoken language (and to regular schooling). This is how its benefit had typically been assessed. In Sweden, the implant's fundamental purpose was different. Combined with sign language use, the cochlear implant was to facilitate bilingualism: that is to say, the deaf child's access to spoken and written Swedish as a second language. If this was its purpose, then almost all foreign research, starting as it did from another premise and focusing on another outcome, was irrelevant to the Swedish situation. The reasoning of the Swedish Board of Health makes sense. In Sweden, the objective in establishing a program of cochlear implantation was to promote bilingualism rather than "make deaf children hearing" and so move them to mainstream schools. If this is the objective, then Preisler's work in Sweden is far more important than most of the two or three hundred papers published annually from centers elsewhere.[9] As to whether that *should* be the objective—the question was rarely posed outside Sweden. Nor could it have been without threatening the complex network of interests in which the cochlear implant had become embedded.

The evidence that, after 1990, helped persuade health policy makers, hospital administrators, clinicians, and parents of deaf children to pursue cochlear implantation was constructed on the basis of a conviction that implantation led to better hearing—and that hearing could stand as a sufficient proxy for all else. That there was more to a (deaf) child's development than this, and that too little was known of these other things, was regularly acknowledged but without giving any significance to that acknowledgment beyond noting it as a future research priority. To forestall any challenge to professional status and jurisdictional authority, the relevance of other knowledge and experience (whether of members of the Deaf community, linguists, educators, or psychologists—all those who might otherwise have usurped or complicated the evaluation of a medical practice) was marginalized.

Today much store is set by evidence, and evidence-based medicine is an ideal for many epidemiologists, health policy makers, and third-party payers. Professional consensus regarding the merit of a test or treatment, and for whom, should be based on rigorous appraisal of the evidence. Advocates of evidence-based medicine acknowledge that many older practices were never evidence-based and that where incentives were strong enough, early adoption of new technologies sometimes ran far ahead of proofs of their value in patient care.[10] Far from being a problem for advocates of evidence-based medicine, acknowledging past inadequacies is crucial in making clear the need for the new and more rigorous approach that evidence-based medicine claims to provide. In the same way, today much effort is put into aggregating results from clinical trials carried out all over the world—so-called meta-analysis. The assumption is that from such a meta-analysis a still more trustworthy, robust measure of the effectiveness of a treatment will emerge: a measure in which clinicians everywhere can have confidence. What this study has shown, however, is that rather than consensus regarding use of the implant in children having been evidence-based, it was evidence that was consensus-based.

The power to define what was to count as evidence for the implant's effectiveness and to respond to external challenges through strategic addition to the evidential base was crucial to its success. So too was sustaining the claim that evidence was universally valid. No matter that studies of drugs, genetic tests, and organ transplantation, among other things, have convinced anthropologists and STS scholars that the meaning and significance of medical technologies vary profoundly from culture to culture.[11] Theirs was no challenge in practice. Professional and institutional convictions, commitments, and interests lent a dynamic to the diffusion and institutionalization of the implant that is well captured by historian Thomas Hughes's notion of "momentum." Hughes developed his concept by reference to large-scale technological systems: electricity generation and supply, interurban transportation, and industrial production.[12] He intended the term to capture the changes that take place in a technological system as social and institutional interests become vested in it, and by virtue of which change becomes increasingly difficult. Though on a more modest scale than Hughes's examples, the history of cochlear implantation shows similar processes.

In the 1990s, implantation practices began to be introduced in poorer countries and regions, a process that is now extending to the very poorest. Professional networks and manufacturing corporations play complementary roles in sustaining this momentum. Because cochlear implantation was, and remains, an innovative and high-status practice within the speciality of ENT (and perhaps more generally), both professional and pecuniary incentives are at play. Surgeons trained in leading American, Australian, or European centers can count on the support of their erstwhile colleagues. These supporting networks, as well as the support provided by the manufacturers, are crucial in making possible the establishment of implantation centers that would cater to rich families with a deaf member. Collaborating centers are provided with technical support (and provide clinical data in return). Manufacturers establish subsidiaries in distant parts of the world, market the technology, and organize associations of users to lobby on their behalf. Inspired by the hope held out to them, elite parents of deaf children become an important force in the spread of implantation. In poorer regions of the world that lack the most basic audiological services for the majority of the hearing-impaired, the beginning of a service like this is greeted as a triumph by the mass media, an important token of the country's modernity.

Languages and Opportunities

With the notion of alternatives introduced in chapter 1, I wanted to bring together a set of disparate activities. Each involves a search for physical and mental well-being that rejects something fundamental to medical practice rooted in a particular epistemology, a professional jurisdiction, or in the consumption of ever more technologically sophisticated services. Theoretical underpinning may lie in a rejection of the reductionist epistemology of biomedicine, as in the environmental breast cancer movement or in CAM. It may be rooted in an aversion to something in the culture of medicine: the aggression and the force with which new technologies are forged or the new forms of subordination and dehumanization to which they sometimes give rise (as in the international organ market). They may be grounded in a critique of the processes by which aspects of being-in-the-world that were once seen as inherent to human variability—such as shyness, aging, childlessness, being of short stature—are being medicalized. They may critique the ways in which the health of the population as a whole is being subordinated to the demands of privileged individuals: in which inequalities in health are exacerbated by technological advance. With the notion of alternatives, I tried to refer to something rooted in practice as well as in social theorizing. I wanted the concept to refer to the articulation in practice of non- or differently medicalized forms of caring. The disability rights movement, using a social model of disability to ground its rights-based demands, was one example. The pro-anas, seeking safe ways of living with, rather than curing, their anorexia through the exchange of experiences, offered another example.

In its early years, few deaf people saw themselves as potential cochlear implant users. Going much further than simply not volunteering as implant candidates, Deaf community activists began to see the implant as symbolizing a century of suppression of sign languages and Deaf culture. Inspired by research in sign language and deaf studies, increasingly self-confident, they began to protest the discrepancy between their own experience and aspirations and the view of deafness endlessly reiterated in support of the cochlear implant. They objected, for example, to the common tendency to equate "language" with "spoken language," as though sign languages did not exist or were not true languages. Protest became far more vehement when, after 1990, implantation was extended from adults to deaf children. Deaf activists began to question the competence of hearing parents, who knew nothing of the lives of deaf people, to make reasoned decisions on behalf of their deaf children. They objected to the assumption (essential to the economic case for pediatric implantation) that a deaf child is better off in a regular mainstream school than in a school for deaf children. The claims being made for the cochlear implant not only showed no knowledge of the lives of deaf people but, far worse, represented yet another assault on the aspirations and the rights of the Deaf community. Deaf advocates crafted a different historical rendering of the implant, one that embedded it in a history not of medical progress but of the oppression of deaf people.

Extrapolation from the life experiences of deaf people who had succeeded in living fulfilling lives was fundamental to the alternative that Deaf advocates were advancing: the claim that a deaf child's interests are better served not by a place at the margins of hearing society but by growing up as a member of the Deaf community.[13] Not only were the results of cochlear implantation limited, certainly when measured in terms of functional participation in the hearing world, but membership in the Deaf community offered the child the vital benefit of a positive sense of self. Since membership in the Deaf community is psychologically and socially beneficial in itself, then even were the device wholly effective in providing hearing, access to this community would still be more desirable. Moreover, there was the risk that, expecting so much from the implant, parents of children newly diagnosed as deaf would not take the trouble to learn to sign. However much or however little the child might ultimately profit from the implant, it would be deprived of access to language for a vital period of its early life, a point that much concerned the French committee on medical ethics.

The lived experience of successful deaf people, and the circumstances that had made their lives possible, could have provided an alternative starting point for assessing many important aspects of the way in which the implant was used. For example, is it better to maintain or avoid the use of sign language (at school, in the home) even when a child has received an implant: better for the child as a whole, not just for its speech? But the experiences of deaf people and the circumstances that had enabled some to be successful and happy were not allowed to inform evaluations of the implant. Considering experience would have posed a threat to established professional and institutional jurisdictions and interests. Like the expertise of linguists and psychologists, it would have challenged the authority of specifically medical expertise. And it would have threatened the broader network of institutional interests and practices within which the implant had become embedded. Many implant programs were closely associated with oral schools for the deaf, and they insisted that parents commit themselves to the exclusive use of spoken language. Through emphasis on the exclusive use of speech at home and attendance at an oral school, the value of the implant was to be maximized.

Appeals to experience having failed to persuade, Deaf advocates turned to bioethics and an argument based on rights. Given bioethics' initial concern with, in Daniel Callahan's words, "the likely effects of biomedical knowledge and its application on the human condition" and with the effects of biomedicine on human welfare, this seemed a reasonable way of legitimating their critique. Bioethics provides the most authoritative language available for debating the broader implications of new medical interventions. Harlan Lane and others thus argued that the Deaf community and Deaf culture had a right to protection and respect. Parents of deaf children, trying to decide what was in their child's best interests, had an obligation to listen to what the Deaf community had to say. This challenge failed too, because advocates failed to appreciate the extent to which bioethics had already been shaped by professional interests and by a broader philosophy of liberal individualism. Initial concerns with the human condition had been set aside as the field crystallized.

In the last few years, a number of ethicists, philosophers, and social scientists have begun to make a case for a more comprehensive bioethics in which these broader concerns are returned to their proper place.[14] One is Harvard ethicist and philosopher Norman Daniels, who has long been concerned with bioethicists' lack of attention for issues of social and distributive justice.[15] Another is Georgetown University law professor Gregg Bloche, who has discussed the circumstances under which physicians typically do subordinate the interests of individual patients to some notion of the greater good.[16] This happens, for example, where public health is at stake. Where rates of immunization are very high and the incidence of a particular disease is low, it is possible that the benefit of vaccination to any individual is outweighed by the risk of some adverse effect associated with the vaccine. The physician nevertheless recommends vaccination because the collective benefit of sustaining high levels of vaccination is deemed the greater good. The medical profession has long-established public responsibilities or social purposes that may conflict with, and may need to be balanced against, the principle of doing one's best for the individual patient. Attaching absolute primacy to "fidelity" to the individual patient is belied by social reality in which conflicts of roles and responsibilities have always had a part. Bloche's argument is not that the demands of the state should always take precedence over the interests of the individual patient, but that sometimes they do, and should.[17]

Ethicist Margaret Olivia Little has developed an argument related to Bloche's. Little is concerned with cosmetic surgery, which seeks to alleviate suffering "which is in some sense due to social attitudes and norms rather than some disease or biological dysfunction."[18] She suggests that among these norms and attitudes are some that can be seen as "part and parcel of an unjust social ideology." What if the surgeon has moral doubts regarding racist or sexist norms to which the patient is trying to adhere? Little suggests that under such circumstances, acting with integrity requires that the surgeon take this broader normative context into account. To reason solely in terms of professional responsibility to the individual patient is not sufficient. Bloche stresses the importance of "an unfailing moral alertness—a habit of self-conscious reflection on the legitimacy of social purposes and on the ethical tensions between public obligation and Hippocratic fidelity." Little's argument leads in the same direction but acknowledges that in addition to repressive state purpose, unjust social ideologies also demand moral alertness on the part of the physician or surgeon.

According to this view, for a doctor to act with integrity means that he or she must take account of the broader normative context. It involves moral reflection, trying to balance the norms and interests sustained by a particular medical practice at a particular time and place against the wishes of the patient. The challenge for the morally alert physician becomes trying to identify what, socially and morally, might be at risk, for example in the expansion of cochlear implantation. The political philosopher Charles Taylor has argued that a social and cultural minority, in which membership is the source of an individual's self-respect, possessing language and culture with which people strongly identify, has to be accorded certain rights.[19] When that group faces discrimination or threat, striking at its members' sense of dignity and worth, a fundamental value is at stake. Whereas principlist bioethics could not provide the theoretical grounding for their case that Deaf advocates were seeking, arguments stressing collective rights and the moral responsibilities of the medical profession can do precisely that.[20]

Deaf advocates thus argued that use of sign language and active membership in the Deaf community provided the deaf child with an alternative and ultimately more rewarding route to adulthood. In practice, associations of deaf people did not always find it easy to associate themselves actively with this view (as for example in the United Kingdom), and in some countries the initiative was taken up by radical splinter groups such as the French Sourds en Colère. Despite support from the World Federation of the Deaf, the alternative view has had relatively little influence in much of the world as pediatric implantation, accompanied by a familiar discourse of normalization and hope, has spread around the globe.

In most countries, Deaf communities' possibilities for arguing their case were limited. As sociologists discussing new social movements have shown, having arguments available is not enough to explain success in influencing (or failing to influence) public opinion or policy. Opportunities to speak being available in one political system or another also have to be taken into account. For example, in the Netherlands the political decision to reimburse pediatric implantation was prepared within a medically dominated advisory body (the Health Insurance Council). Its advice tended to be followed automatically by the minister of health. Even though at the same time a separate committee to advise on the implications of official recognition of Dutch Sign Language had been appointed by the same ministry, that ministry was unable to see that the issues were related. Through behind-the-scenes lobbying, medical professionals and academic teaching hospitals were able to force the minister's hand. By contrast, the Deaf community lacks representation in institutionalized advisory structures and it lacks channels of informal political influence almost everywhere.[21] Looking at how the alternatives appear to parents of newly diagnosed deaf children, comparable inequalities appear. Even before implantation became as "natural" as it has become in some countries, the recommendation to seek an implant was made in a carefully cultivated relationship of trust that few parents questioned. By contrast, what the Deaf community had to say was unknown to many parents, came from a marginal or stigmatized source, and was not easy to access even for the few who tried. A core element of their message—the notion of a "Deaf community," of which membership is said to be a "birthright" and a source of identity—was not an easy one for parents to accept.

The alternative approach envisaged by Deaf community advocates has acquired nothing of the momentum that the implant has acquired. No institutional interests have been mobilized behind it. There is an additional impediment to enacting an alternative. The existence of a shared language, of schools and other institutions for the socialization of deaf children and the promotion of deaf sociality, the existence of a cohesive Deaf community, are what give local meaning to the possibility of a Deaf identity. For many deaf people, in much of the world, a positive identity as Deaf is simply not available. The alternative that Deaf advocates hold out, bringing up a child as a member of the signing Deaf community, is contingent on the existence of such a community. However, in many countries a national Deaf community scarcely exists, and deaf people remain isolated and stigmatized. Recent research has begun to throw light on the relatively recent formation of Deaf communities in some countries and the social and political changes that catalyzed their formation. For example, in what is now Russia, the reforms introduced by Gorbachev were fundamental; sign language schooling for deaf children only began in 1992. It is still neither widespread nor widely accepted in the country.[22] The concept of "Deaf community" is as yet scarcely known, and there is no established term for it in the (spoken) Russian language. The same holds for "deaf culture" and "deaf identity."[23] In Nicaragua until some thirty years ago, being deaf almost inevitably meant a life of isolation. Richard Senghas writes, "Before 1978, there was no established Deaf community in Nicaragua; older deaf people had no ways to pass down the wisdom of deaf experience or to tell stories of the old days. There was no shared sign language."[24] It was only after the Sandinista revolution of 1979 and the foundation of the first residential schools for deaf children in the early 1980s that a national sign language and a Nicaraguan Deaf community came into being. In many countries, comparable developments have still to take place.

The idea that a cochlear implant and a Deaf identity stand for alternative approaches to bringing up a deaf child, at the root of the technology's controversial history, is true at the symbolic level. Only under certain conditions are they alternatives in practice. Illustrative of the perception that the meanings of technologies are mutable, common to both anthropology and STS, it is possible to see how and why the cochlear implant can without contradiction be at once hope, threat, and irrelevance. But meanings are also in flux. In much of Asia, Latin America, and Africa, processes of market-driven globalization are bringing the cochlear implant within reach of a privileged minority of deaf children. But processes of social and political reform are bringing the possibility of a Deaf identity within reach of the less privileged majority.

Mechanisms and Processes

Where decisions are made within a medical arena and (ostensibly) on medical grounds, then the quality of the evidence becomes crucial. Rigorous and robust trial data, taking costs and QALYs or DALYs (disability adjusted life years) into account, should provide justification enough. It was on the basis of such reasoning that the Dutch pediatric implantation program submitted the report that it did submit, excluding virtually any reference to controversy and areas of ignorance. Using the same reasoning, the Health Insurance Council decided that the value of the implant had been proven. But this depends on decision making being "contained" in an expert, medical forum and not "leaking out" into politics or the courts. Knowing that the procedure was controversial and perhaps concerned by its cost implications, the Dutch minister of health, unusually, hesitated at adopting the council's advice. She decided to consult the various stakeholders, thereby coming into conflict with the Health Insurance Council who considered the case for implantation to have been proven. Situations like this are becoming more common: a result of growing demands for wider participation and transparency and declining trust in experts. What happens when an expert forum loses its legitimacy: when its approval or guarantee cannot reckon on widespread trust?[25]

One possibility is that its membership will be broadened or its procedures opened up. Another possibility is that controversial issues will be referred to some other forum in the system of governance. Charged with representing the public interest, with no prior commitment to medical evidence or reasoning, obliged to gain the trust of the public, such a forum is an unknown quantity for the experts who have lost control of the issue. For a host of regulatory bodies in the medical and health fields gaining the trust of the public is crucial: the key test of their legitimacy. Regulatory bodies in fields like genetics or stem cell research, where there is considerable controversy, have been forced to accommodate the wide range of conflicting perspectives set out by patient groups, religious groups, industrial associations and scientists. In order to do so, seeking to establish the legitimacy of their conclusions, they have broadened their memberships to include members of the public, and they have developed consensus conferences and other consultative mechanisms. Legitimacy, in other words, is to be derived from a process of democratic deliberation.[26]

Over the past decade or so, the subject of pediatric implantation has been taken up by a number of forums that, on the face of it, appear to be concerned with reconciling the divergent views. However, there are big differences in how each of them came to take up the subject, as well as in their functioning and standing. At different places in this book I have referred to three such forums, and it is instructive now to look at them together.

NIH CONSENSUS DEVELOPMENT CONFERENCE OF 1995

NIH consensus development conferences have become a standard element of NIH practice and are organized according to an established procedure. An independent panel of experts is appointed to prepare a consensus statement. It meets for a day or two in public, hears presentations by investigators in the field, and discusses questions raised by the audience. As discussed in chapter 2, in May 1995 NIH convened a second consensus development conference on the subject of cochlear implants. Of the fourteen-member expert panel, ten were specialists in otolaryngology, hearing science/audiology, and biomedical engineering.[27] The resulting consensus document notes, "The conference was convened to summarize current knowledge about the range of benefits and limitations of cochlear implantation that have accrued to date. Such knowledge is an important basis for informed choices for individuals and their families whose philosophy of communication is dedicated to spoken discourse."[28] The introduction explains, "Issues relating to the acquisition of sign language were not directly addressed by the panel, because the focus of the conference was on new information on cochlear implantation technology and its use." Having limited itself to the information needs of individuals and families already wholly committed to oralism and to the question of how benefit from the implant can be maximized, this conference avoided the controversy totally. As a result, it had little difficulty in agreeing on a position that wholly reflected medical thinking at the time.

FRENCH NATIONAL CONSULTATIVE ETHICS COMMITTEE ON HEALTH AND LIFE SCIENCES' "OPINION," 1994

Established in 1983 by the president of the French Republic, the National Consultative Ethics Committee on Health and Life Sciences (CCNE) is a consultative body with a broad mandate. Its mission, defined by law, is "to give opinions on ethical problems and societal issues raised by progress in the fields of biology, medicine and health." The committee takes up issues "referred" to it by the country's president, by a member of the government, or by a university or other public institution working in the research or health fields. In addition, if it so decides, the committee may take up an issue referred to it by a private individual or one of its own members.[29] The subject of pediatric implantation was referred to the CCNE by a radical movement (Sourds en Colère) together with a group of psychologists, sociologists, linguists, and educators. This group invited the committee to consider whether, given the uncertainties surrounding the social, psychological, and linguistic implications of implanting children, the practice should be ruled as experimental under French law. The committee agreed to take the issue up and issued its opinion at the end of 1994. As discussed in chapter 4, although the committee rejected the claim that the procedure should be defined as experimental (on the ground that practice was too far institutionalized), it did conclude that, to avoid the risk of compromising children's social and psychological development, all deaf children should be offered sign language from an early age, whether or not they would later be candidates for implantation.

DUTCH PLATFORM, 1995–1999

The Dutch Platform, an ad hoc consultative forum, emerged from a two-day meeting at which representatives of the two implant teams, parents of deaf children, and members of the Deaf community had debated the issues raised. The platform that was then established under an independent chair contained representatives of each group: a form of deliberation that is well established in the Netherlands. As discussed in chapter 5, the platform initially functioned well as a place at which differences of opinion could be debated. The value of long-term research on the effects of implantation, for example, was an issue on which agreement was found. However, it turned out that consensus was possible only as long as nothing was at stake. When the search for consensus appeared to be holding up ministerial approval of reimbursement in 1997, unbridgeable tensions emerged, and when approval was obtained in 1999, the implant teams were no longer interested in participating in the platform and it collapsed.

Like the Dutch surgeon who, early in the history of implantation, explained that he planned simply to "ignore the opposition," the NIH consensus development conference defined "consensus" so narrowly as to exclude all matters of controversy from its remit. The French National Ethics committee came to a conclusion that pleased the French Deaf community but not the medical profession. However, its report had little, if any, influence on the subsequent course of events: a clear example of ethicists being unable to "hold the line." Though lacking any official standing, the Dutch Platform did bring representatives of all stakeholders together over a considerable period of time. While it seems to approach most clearly the deliberative democratic ideal, a precondition for legitimacy in the regulatory field, it too failed.

In trying to draw lessons from these three failed attempts at seeking a workable and acceptable consensus, political theorist Iris Marion Young's discussion of the limits of deliberative democracy is helpful.[30] Young frames her discussion as a dialogue between proponents of two competing positions in the struggle for social justice. The "deliberative democrat," as represented by Young, "thinks that the best way to limit political domination and the naked imposition of partisan interest and to promote greater social justice through public policy is to foster the creation of sites and processes of deliberation among diverse and disagreeing elements of the polity." He (or she) claims that that through "reasonable argument," policy acceptable to all can be agreed. The "activist" mistrusts the invitation to participate in such a deliberative process, however, because deep-rooted structural inequalities bias the rules of any such process and constrain possible outcomes. It therefore makes no sense to "sit down with those whom he criticizes and whose policies he opposes to work out an agreement through reasoned argument they can all accept. The powerful officials have no motive to sit down with him, and even if they did agree to deliberate, they would have the power unfairly to steer the course of the discussion."

The activist critic of deliberative democracy, in Young's discussion, has four criticisms. Deliberative processes are exclusive, with rules, participation, and the options to be considered determined by those in power. The activist should be outside, like the antiglobalist protestors at World Trade Organization meetings, objecting to the exclusiveness of the process. Deliberative democrats have understood and to a considerable degree accommodated this criticism, says Young. Members of the public are invited to testify. But the activist remains suspicious, since however "open" the deliberative process is made, under conditions of structural inequality access will in practice still be restricted to those with sufficient knowledge and resources. The proper place for the activist is still outside, speaking on behalf of those who remain excluded. The democrat retorts that compensatory measures can be found that enable broader and more representative participation if only the activist would come in and make clear what needs to be done. Granted all this, the activist also believes that social and economic inequalities constrain the range of options that are open to discussion. Given the way options are set and foreclosed, participation only serves to give legitimacy to the structures and the relations that have set the agendas and fixed the options. Young's view is that to the extent that alternatives in a deliberative process are fixed by institutional priorities and social structures, themselves the source of inequality and that remain nonnegotiable, "deliberation is as likely to reinforce injustice as to undermine it." The fourth area of challenge has to do with the way in which discussion will proceed, the "discourse": the ways of thinking and reasoning that frame debate and that shape how members of a society think about social relations and processes, whatever their position in society. "Hegemonic discourse" "refers to how the conceptual and normative framework of the members of a society is deeply influenced by premises and terms of discourse that make it difficult to think critically about aspects of their social relations or alternative possibilities of institutionalization and action." Because of limits on the ways we can think about and discuss things (even though we are unaware of those limits), there is a real likelihood that "false consensus" might emerge. The activist, then, is enjoined to "continue to challenge these discourses and the deliberative processes that rely on them, and often he must do so by nondiscursive means—pictures, song, poetic imagery, and expressions of mockery and longing performed in rowdy and even playful ways aimed not at commanding assent but disturbing complacency."

The "activist" will surely feel his or her position vindicated by the American, French, and Dutch deliberations of cochlear implantation in children. The first three of Young's arguments are sufficient to explain both the self-limitation imposed by the NIH procedure and the collapse of the Dutch platform. The French case, the inability of so respected an ethical body to "hold the line" in practice, provides the activist with yet a further argument beyond those discussed by Young.

The existence of some common ground, some measure of mutual understanding, is a precondition for constructive deliberation to take place. The attempt to find such common ground, to establish such mutual understanding, is inherent in the "mediating and intervening" position proposed by anthropologists Downey and Dumit. In chapter 5, I explained how I came to adopt a research strategy like that, trying to promote dialogue, mutual respect, and a course of action to which all sides could subscribe. It also entailed a growing appreciation of the fact that public conflict, essential though it was to the Deaf community's struggle to achieve full citizenship, was also a major source of anxiety for parents. The fact that the cochlear implant had been vested with such symbolic weight—had become the locus of a struggle over both professional authority and social inclusion, dignity, and respect—had a different significance for the parents of deaf children. For them—for us—it was (is) a promising intervention regarding which we have to try to make an informed decision.[31]

In different ways and for different reasons, attempts at resolving the controversy in principle, whether through adjudication by experts, on the basis of stakeholder deliberation, or through mediation, all failed. Similarly, the vast bulk of writing on the implant controversy and most of the research that has been carried out has been in support of one position or the other. Dr. Bredberg, the Swedish surgeon quoted in chapter 4, pointed out that "the problem of the cochlear implant is very much a pedagogical problem." His view is not reflected in the research literature, most of which is concerned with technique, surgery, or audition. With the possible exception of Sweden, neither proponents of pediatric implantation nor those opposing it have been willing to base their claims—what implantation offers, to whom, and under what conditions—on the results of empirical investigation.

Yet, slowly but surely, a more nuanced picture of what the cochlear implant does in practice *is* emerging. Some adult deaf people with an implant tell of their lives in ways that show something much more complex than the "either/or" controversy allows for. The stories told by Beverly Biderman in Canada and by Nicole in France show that for them it was not a matter of choosing for the Deaf or the hearing worlds. It was a matter of moving between the two, of negotiating a "way of being deaf," that suited their circumstances and their life choices. For each of them, the implant was a vital aid in living the lives—as Deaf people who could hear and speak when they chose to—that they wanted to live. In doing so, in retaining their Deaf identities but functioning also in the hearing world, they were doing something that should seem wholly unsurprising. It is no more and no less than countless young people born into ethnic or religious minority communities also do: seeking ways of adapting to and functioning in the wider world while giving their own meaning to their origins and the community in which they were brought up. Gallaudet sociologists Christiansen and Leigh were surprised to find that the vast majority of American parents they surveyed said that, as far as they were concerned, "their child was still deaf after implantation." Writing in 2002, their view was that "the walls between those who support pediatric implantation and those who oppose the procedure are, if not crumbling, at least beginning to crack."[32] Despite all the failures to find an acceptable and workable compromise in principle, and however reluctantly, each party is at least starting to listen to what the other has to say. True perhaps of the United States and possibly of parts of Europe, but it is hard to see how so subtle and complex a view, the product of particular social configurations, is likely to reach those more distant countries in which parents of deaf children are now pinning so much hope on a technology that few will be able to afford.

Lessons to Be Learned

For me personally, the cochlear implant is not simply one among a myriad of new medical technologies. Once, like many parents of deaf children, I pinned so much hope on it. And because my sons Jascha and Boaz are very much part of the Deaf community, any technology or practice that could significantly affect the future of that community still has great emotional significance. One consequence was that carrying out this study, over many years, was not easy. It was not easy to "find the mean," the right degree of emotional detachment. Sometimes an audiologist or a surgeon whom I was interviewing tried to persuade me that in rejecting an implant for my own child I had denied him something of great value. That was painful. Attending conferences on pediatric implantation could have a similar effect. I sat through dozens of presentations by surgeons, audiologists, and speech therapists. The more "scientific" of them, illustrated with slides of data taken from tens or hundreds of children, full of correlation coefficients and measures of statistical confidence, didn't affect me personally. I responded to them as a sociologist, as I would to any other scientific presentation. But others, illustrated with "before and after" videos of an individual child showing how well the child had learned to speak, often left me sad and unsure. Television programs about cochlear implantation, featuring individual children and their parents, sometimes had the same effect. I saw the uncertainty, but also the hope, of parents like myself. Early in my work, unused to such programs, I was caught up in the drama. Overcome by emotion, I wasn't able to change register. To do my sociological work meant distancing myself from the televised parents and children. It meant asking questions like "how has this program been put together?" "What message is it trying to convey?" "Why is it being shown now?" With time and experience, it became easier. I learned to see the programs as having a characteristic message and a standard plot. By providing an alternative vantage point, my sociological inquiry provided an emotional protection that is not available to most parents of deaf children.

Despite my emotional involvement with this technology, I believe that important lessons of much wider significance can be drawn from this account of its development and the controversy around it.

The first concerns the terms in which the benefits of new medical technologies are to be assessed. Consensus regarding the utility of the implant, first in adults and then in children, was based on evidence for its effect on hearing. From time to time, it was acknowledged that the consequences of implantation for children's development were far broader than that. But in the period in which professional consensus was formed, virtually no studies of the effects of implantation on deaf children's linguistic, cognitive, or socioemotional development had been conducted. There had been no studies of the consequences of early implantation for parent-child relationships. There had been no studies of the benefits (or drawbacks) to the child of being moved into mainstream education (a move so vital to the economic case for pediatric implantation). Finally, and in striking discord with today's talk of "experience-based knowledge," the knowledge and experience of the Deaf community was not allowed to play any part in assessing the benefits of cochlear implantation.

In chapter 1, I suggested that two competing trends are at play, each of which bases it claims to legitimacy on a particular notion of patient empowerment. One of these is a consumerist notion, the other a more radical and critical one. The claim that the Deaf community possessed the appropriate experiential knowledge was countered by establishing "cochlear implant users' associations." The question then is why and how was so much excluded from assessment of the implant? "Why," according to this analysis, is because to have done otherwise, to have made professional consensus dependent on the results of psychosocial and linguistic studies (let alone on the experiences of deaf people), would have been to compromise professional status and jurisdictional authority. Deaf advocates were posing a challenge to medicine. They were questioning the competence, the authority, and the right of the profession to assess an intervention that it defined as medical in its own restricted terms. The proper test was not "improved hearing," measured with audiological scales and instruments, but a far more complex metric rooted in psychology, linguistics, and the reflexive understanding of deaf people. However, hearing and its measurement—and these alone—lay within the undisputed jurisdiction of the medical and audiological professions. "How" it happened was because, other than in Sweden, neither Deaf communities nor disciplines such as psychology and linguistics were allowed more than a modest and subordinate role. The first general lesson that this study has to teach thus concerns how the terms in which the benefits of a new intervention are to be established should themselves be determined, and who should be involved in their determination. In practice, these questions are almost never raised. In a systematic review of the establishment of outcome measures in clinical trials involving children, a group of pediatricians and statisticians from the University of Liverpool in England found that the question of "what to measure" was rarely posed. Moreover, of the studies that did consider the matter, few involved parents and none involved children in determining the appropriate end points for clinical trials. They conclude, "It seems logical that their involvement would help determine the most appropriate outcomes to measure, but there is no evidence to substantiate this hypothesis, nor is there a framework that would recommend the best strategy for involvement."[33]

The second set of issues has to do with the inability of the Deaf community to gain serious attention for its point of view (again, other than in Sweden). What is involved here goes further than lack of participation of deaf people in the initial evaluation of the technology. Beyond arguing that a Deaf identity offered more to the deaf child, in terms of self-esteem, than a place at the margin of hearing society, the Deaf community also argued that its own well-being, its future, was threatened by the growth of pediatric implantation. Its inability to have this concern taken seriously was due in part to the relative weakness of channels through which its views could be diffused and in part to its lack of voice in national structures of political deliberation. This was (and is) all the more true of those many countries in which the Deaf community is either fragmented or scarcely organized and where, as a result, a positive Deaf identity is not available. But it was true also of countries such as France, the Netherlands, the United Kingdom, and the United States. Bioethics, which seemed to offer an authoritative language and a forum for debating the anxieties of the Deaf community, did not do so because of the way in which its scope had become restricted to matters of individual rights. Rethinking the scope of bioethics, reintroducing broader concerns with the responsibilities of the medical profession to society as a whole, issues of social justice, collective rights, the implications of medical advance for human flourishing broadly conceived, in ways proposed by Daniel Callahan, Renée Fox and Judith Swazey, Norman Daniels and other writers quoted in this book, offers part of the solution. Fox and Swazey, participant-observers in the bioethical enterprise since its inception, and many (though not all) the ethicists they interviewed for their recent book see signs of the emergence of this richer and more culturally sensitive bioethics.

In the meantime, in some countries workable ways forward that neither expert nor deliberative political processes could deliver are now emerging. The implications of the fact that compromise comes not from political decision making but from civil society has relevance for how controversial medical technologies more generally are to be dealt with. How can a decision-making system do justice both to a political claim (such as that of the Deaf community that it has rights that merit respect and protection, and that are now threatened) and to the conviction of individual citizens (such as parents of deaf children) that a treatment is in their or their children's personal interests? Perhaps a way forward follows from the argument formulated by social theorists Alberto Melucci and Leonardo Avritzer: "If democracy is to keep its legitimacy, it needs to assume a different form in complex, pluralistic societies," they write. "It has to create a space for solidarity and public presentation within which social actors recognize themselves and can be recognized for what they are or want to be."[34] They suggest that this implies "broadening the concept of democracy" to include "the freedom to construct spaces for recognition, the freedom to dispute the meaning of given identities and the freedom to innovate at the political level." What this might mean in practice was hinted at by Rayna Rapp and Faye Ginsburg. Their argument regarding disability, which applies equally here, was that the integration of disability into everyday life, the shift from exclusion to inclusion, has to be rooted in family life. "Public storytelling," testifying to the ways in which families can accommodate children with special needs, is an essential resource in enabling parents to do the necessary imagination work.

The third set of issues relates to what, adopting Thomas Hughes's term, I called the "momentum" of the technology. This momentum has its origins early in the history of cochlear implantation, as ENT surgeons, industrial corporations, and institutions and professions providing services to the deaf came to see it as a source of status, legitimacy, and profit. Drawing on familiar and culturally powerful notions of medical progress, they had little difficulty in mobilizing the mass media behind their view of the implant as a source of hope for parents of deaf children. Parents, their hopes aroused, provided not only a receptive market but also a political resource. Appeal to the length of waiting lists, to the notion that children's future benefits would be compromised by waiting, was an argument that could be used politically to free up health care resources. A consequence has been that in many western countries, parents now choose to have their deaf child implanted not after careful weighing up of alternatives but on the basis of unthinking emulation. Despite the doctrine of informed consent to which all now play lip service, the social dynamic of momentum leads inevitably to a situation that excludes precisely the informed and reasoned choice that the doctrine requires. It is this social dynamic that lies behind the problem that ethicist William Gardner identified: the inability of ethicists to "hold the line."

In the course of the last decade, cochlear implantation has spread to middle-income and to poor countries. Professional status considerations, collaborative networks, marketing and support provided by manufacturers—in short, the momentum of the system—propels the technology into poor countries lacking even the most basic provisions for deaf people. Though Chinese, Indian, and Korean scientists may succeed in developing implants costing one-quarter or one-fifth as much, its present cost of $20,000, plus the costs of prolonged rehabilitation, puts it out of reach of all but the very wealthiest citizens of poor countries. Exacerbating inequality, its appropriateness in countries in which few hearing-impaired people can afford a simple hearing aid and in which most deaf children receive little or no schooling is surely questionable. In resource-poor settings, the cochlear implant can only aid a relatively privileged minority at the expense of the doubly disadvantaged majority of deaf people. "At the expense of," because the sheer fact of the implant will be used to justify continued failure to make the social provisions that, as in Nicaragua, led to the emergence of a national Deaf community and created the possibility of a Deaf identity.

The fourth and final lesson to be drawn from this study is rather different from the first three. It concerns our own scholarly interrogations of the world of medicine and illness.

Responding to the number of health-related social movements that have emerged in the last twenty years, to their successes, and to the significance of health in the modern world, historians and social scientists have shown increasing interest in patient activism.[35] Studies have described the struggles that have been fought for the acceptance of patients' experience as knowledge, "different from professional knowledge but essential for understanding and improving their condition."[36] Most published accounts of patient activism describe success, or something approaching it. They show how patient groups have been able to influence research agendas, the direction of research, and the accessibility of new therapies. They thus serve as exemplars, as a valuable source of inspiration, for other associations of patients, other "biosocial" groupings, striving for comparable acknowledgement and influence. They in effect bear witness to the validity of the claim that patients today are indeed "empowered." By providing concrete empirical examples, they suggest that currently fashionable notions of patient empowerment and the democratization of research are more than empty slogans.

The analysis offered in this book has been very different. The demands of Deaf community leaders and advocates had little or no effect either on development of the cochlear implant or on the beginnings of local implantation practices. The experience of deaf people was not accepted as essential or even as relevant. Just as Young's activist critic scrutinizes the real workings of deliberative democracy, so too the extent, limits, and distortions of patient empowerment have to be critically examined. The significance of the Deaf community's failure to influence the development and spread of cochlear implantation lies here. It points to the limits imposed on empowerment and suggests that patient groups can gain acknowledgment and influence only insofar as their demands are compatible with certain fundamental assumptions of medical science, medical authority, and the consumption of medical goods and services.[37] A failure to point to these limits is a failure to reflect sufficiently on the extent to which we too, historians and social scientists, are influenced by, not to say complicit with, the ways of thinking and reasoning that frame contemporary social and political debate.

Notes

Chapter 1 **The Promise of New Medical Technology**

1. McCay Vernon, "Parental Reactions to Birth-Defective Children," *Postgraduate Medicine* 65 (1979): 183–89.
2. Pauline Shaw, *The Deaf Can Speak* (London: Faber and Faber, 1985).
3. Jessica Rees, *Sing a Song of Silence* (London: Futura, 1984).
4. Ibid., 87.
5. Ibid., 88.
6. David Wright, *Deafness: A Personal Account* (London: Faber and Faber, 1990).
7. Poem "On a Fiftieth Birthday," in David Wright, *Poems and Versions* (Manchester: Carcanet Press, 1992).
8. Harlan Lane, *When the Mind Hears* (New York: Random House, 1984).
9. Wright, *Deafness*, xiv.
10. W. H. McKellin, "Hearing Impaired Families: The Social Ecology of Hearing Loss," *Social Science and Medicine* 40 (1995): 1469–80.
11. Thomas S. Spradley and James P. Spradley, *Deaf Like Me* (Washington, DC: Gallaudet University Press, 1985).
12. The *Volta Review* is the journal of the Alexander Graham Bell Association for the Deaf and Hard of Hearing, a Washington, DC-based organization devoted to the promotion of oral communication among deaf people.
13. Spencer and Louise Tracy's son John was diagnosed as deaf in 1925. In 1942, they established the John Tracy Clinic. It provides distance learning programs designed to help parents of deaf children teach their children to speak.
14. Anonymous, "Dove vrouw 'stomverbaasd' weer geluiden te horen na implantatie," *NRC*, March 14, 1989.
15. "Professor Smit" is a pseudonym.
16. Courtney Everts Mykytyn, "Anti-aging Medicine: A Patient/Practitioner Movement to Redefine Aging," *Social Science and Medicine* 62 (2006): 643–53.
17. Jordan Goodman, "Pharmaceutical Industry," in Roger Cooter and John Pickstone, eds., *Medicine in the Twentieth Century* (Amsterdam: Harwood Academic Publishers, 2000), 141–54.
18. Stuart S. Blume, "Medicine, Technology, and Industry," in Cooter and Pickstone, *Medicine in the Twentieth Century*, 171–85.
19. Stuart S. Blume, *Insight and Industry: The Dynamics of Technological Change in Medicine* (Cambridge, MA: MIT Press, 1992).
20. Stanley Metcalfe and John Pickstone, "Replacing Hips and Lenses: Surgery, Industry and Innovation in Post-war Britain," in Andrew Webster, ed., *New*

Technologies in Health Care: Challenge, Change, and Innovation (London: Palgrave Macmillan, 2006), 146–60.

21. Mary Guyatt, "Better Legs: Artificial Legs for British Veterans of the First World War," *Journal of Design History* 14 (2001): 307–25.
22. Blume, *Insight and Industry*, chapter 3.
23. Heidi M. Bervern and Peter D. Blanck, "Assistive Technology, Patenting Trends and the Americans with Disabilities Act," *Behavioral Sciences and the Law* 17 (1999): 47–71.
24. Alan M. Garber, "The Price of Growth in the Medical-Device Industry," *New England Journal of Medicine* 355 (2006): 337–39.
25. João Biehl, "Pharmaceuticalization: AIDS Treatment and Global Health Politics," *Anthropological Quarterly* 80 (2007): 1083–1126. See also Vinh-Kim Nguyen, "Antiretroviral Globalism, Biopolitics, and Therapeutic Citizenship," in A. Ong, ed., *Global Assemblages: Technology, Politics, and Ethics as Anthropological Problems* (Oxford: Blackwell, 2005).
26. See, for example, Nelly Oudshoorn and Trevor Pinch, eds., *How Users Matter: The Co-construction of Users and Technology* (Cambridge, MA: MIT Press, 2003).
27. Linda F. Hogle, "Chemoprevention for Healthy Women: Harbinger of Things to Come?" *Health* 5 (2001): 311–33.
28. Linda F. Hogle, "Claims and Disclaimers: Whose Expertise Counts?" *Medical Anthropology* 21 (2002): 275–306. A score of 1.7 on one widely used risk-assessment model is taken as the benchmark for high risk.
29. N. J. Fox, K. J. Ward, and A. J. O'Rourke, "The 'Expert Patient': Empowerment or Medical Dominance? The Case of Weight Loss, Pharmaceutical Drugs, and the Internet," *Social Science and Medicine* 60 (2005): 1299–1309.
30. Madeleine Akrich, "The De-scription of Technical Objects," in Wiebe Bijker and John Law, eds., *Shaping Technology/Building Society* (Cambridge, MA: MIT Press, 1992), 205–23.
31. Akrich has suggested that any such discrepancy is crucial to the life of a technology and that "mechanisms of adjustment (or failure to adjust) between the user, as imagined by the designer, and the real user" must be central to sociological research on technological change (Akrich, "De-scription").
32. Julie Anderson, Francis Neary, and John V. Pickstone, *Surgeons, Manufacturers, and Patients: A Transatlantic History of Total Hip Replacement* (London: Palgrave Macmillan, 2007).
33. Stephen von Tetzchner, "The Short History of Text Telephones," in S. von Tetzchner, ed., *Issues in Telecommunications and Disability* (Luxembourg: European Commmission, 1991), 239–44.
34. Joseph P. Shapiro, *No Pity: People with Disabilities Forging a New Civil Rights Movement* (New York: Three Rivers Press, 1993), 214.
35. Susan Reynolds Whyte and Herbert Muyinda, "Wheels and New Legs: Mobilization in Uganda," in Benedicte Ingstad and Susan Reynolds Whyte, eds., *Disability in Local and Global Worlds* (Berkeley: University of California Press, 2007), 287–310.
36. Pandora Pound et al., "Resisting Medicines: A Synthesis of Qualitative Studies of Medicine Taking," *Social Science and Medicine* 61 (2005): 133–55.
37. Rayna Rapp, "Refusing Prenatal Diagnosis: The Meaning of Bioscience in a Multicultural World," *Science Technology and Human Values* 23 (1998): 45–70.
38. Rapp, "Refusing Prenatal Diagnosis," 50.
39. See Phil Brown, S. Zavestoski, S. McCormick, B. Mayer, R. Morello-Frosch, and R. Gasior Altman, "Embodied Health Movements: New Approaches to Social Movements in Health," *Sociology of Health and Illness* 26 (2004): 50–80. These

authors use the term "embodied health movements" in order to emphasize the significance of members' embodied illness experience for their concerns and actions.

40. One influential example is PXE International, which funds, organizes, and conducts research into the rare genetic disease pseudoaxonthoma elasticum (PXE), which causes vision, skin, and arterial defects. Founded in 1995, PXE International is "a small research foundation that is funded by the donations of many individuals and run by advocates, with the purpose of supporting people with PXE and educating clinicians about the condition. The organization has also achieved success in accelerating translational research into PXE, and can be used as a model for other such organizations." See Sharon F. Terry, P. F. Terry, K. A. Rauen, J. Uitto, and L. G. Bercovitch, "Advocacy Groups as Research Organizations: The PXE International Example," *Nature Reviews-Genetics* 8 (2007): 157–64.
41. Mary K. Anglin, "Working from the Inside Out: Implications of Breast Cancer Activism for Biomedical Policies and Practices," *Social Science and Medicine* 44 (1997): 1403–15.
42. Vololona Rabeharisoa and Michel Callon, "The Involvement of Patients' Associations in Research," *International Social Science Journal* 54 (2002): 57–65; Vololona Rabeharisoa, "From Representation to Mediation: The Shaping of Collective Mobilization on Muscular Dystrophy in France," *Social Science and Medicine* 62 (2006): 564–76.
43. For a discussion of patient expertise in fitting a prosthetic device, see Steven Kurzman, "'There's No Language for This': Communication and Alignment in Contemporary Prosthetics," in Katherine Ott, David Serlin and Stephen Mihm, eds., *Artificial Parts, Practical Lives: Modern Histories of Prosthetics* (New York: New York University Press, 2002), 227–46.
44. Fox, Ward, and O'Rourke, "The Expert Patient," 9.
45. M. Klawiter, "Racing for the Cure, Walking Women, and Toxic Touring: Mapping Cultures of Action within the Bay Area Terrain of Breast Cancer," *Social Problems* 46 (1999): 104–26.
46. Sobitha Parthasarathy found considerable differences in the responses of breast cancer activist groups to the introduction of genetic screening in the UK and the United States. In the former the service was provided under the National Health Service (and was widely trusted), whereas in the United States, it was a commercial enterprise. See Shobita Parthasarathy, *Building Genetic Medicine: Breast Cancer, Technology, and the Comparative Politics of Health Care* (Cambridge, MA: MIT Press, 2007).
47. Hogle, "Chemoprevention," 322.
48. Herman Broers, *Dokter Kolff: Kunstenaar in Hart en Nieren* (Amsterdam: Mets and Schilt, 2003), 142.
49. Harold M. Schmeck, *The Semi-artificial Man* (New York: Walker, 1965), 9.
50. Louise B. Russell, *Technology in Hospitals: Medical Advances and Their Diffusion* (Washington, DC: Brookings Institution, 1979), 1.
51. Paul Farmer, *Infections and Inequalities* (Berkeley: University of California Press, 1999), 263–64.
52. "Your Bionic Future," special issue of *Scientific American* 10, no. 3 (Fall 1999).
53. See for example, David J. Mooney and Antonios G. Mikos, "Growing New Organs," *Scientific American* 10, no. 3 (Fall 1999): 10–15; Alex Faulkner, Julie Kent, Ingrid Geesink, et al., "Purity and the Dangers of Regenerative Medicine: Regulatory Innovation of Human Tissue-Engineered Technology," *Social Science and Medicine* 63 (2006): 2277–88.

54. J. R. Ravetz, *Scientific Knowledge and Its Social Problems* (Oxford: Oxford University Press, 1971).
55. Daniel Sarewitz, *Frontiers of Illusion: Science, Technology, and the Politics of Progress* (Philadelphia: Temple University Press, 1996), 17–29.
56. Richard E. Brown, *Rockefeller Medicine Men: Medicine and Capitalism in America* (Berkeley: University of California Press, 1979); Ivan Illitch, *Limits to Medicine* (Harmondsworth: Penguin Books, 1977).
57. Thomas McKeown, *The Role of Medicine: Dream, Mirage, or Nemesis?* (London: Nuffield Provincial Hospitals Trust, 1976).
58. Russell, *Technology in Hospitals*, 1.
59. See, for example, Rayna Rapp, *Testing Women, Testing the Fetus: The Social Impact of Amniocentesis in America* (New York: Routledge, 1999).
60. Renée C. Fox and Judith P. Swazey, *Spare Parts: Organ Replacement in American Society* (New York: Oxford University Press, 1992), 206.
61. Daniel Callahan, *False Hopes: Overcoming the Obstacles to a Sustainable, Affordable Medicine* (New Brunswick, NJ: Rutgers University Press, 1999), 57.
62. Nikolas Rose, "The Politics of Life Itself," *Theory, Culture, and Society* 18 (2001): 1–30.
63. Eliot Freidson, *Profession of Medicine: A Study of the Sociology of Applied Knowledge* (New York: Harper and Row, 1970).
64. Paul Rabinow, *Essays on the Anthropology of Reason* (Princeton, NJ: Princeton University Press, 1996), especially chapter 5.
65. Rose, "Politics," 19.
66. Leprosy (Hansen's disease) offers the clearest and best-described example. See Ronald Barrett, "Self-mortification and the Stigma of Leprosy in Northern India," *Medical Anthropology Quarterly* 19 (2005): 216–30.
67. See Richard C. Scotch, "Disability as the Basis for a Social Movement: Advocacy and the Politics of Definition," *Journal of Social Issues* 44 (1988): 159–72. For the case of Japan, see Reiko Hayashi and Masako Okuhira, "The Disability Rights Movement in Japan: Past, Present, and Future," *Disability and Society* 16 (2001): 855–69. According to disability studies scholar Tom Shakespeare, whereas American disability activism has stressed the extension of rights to inclusion, the British movement has stressed "changing the system that produces disability." Tom Shakespeare, "Disabled People's Self-organization: A New Social Movement?" *Disability, Handicap, and Society* 8 (1993): 249–64.
68. Scotch, "Disability"; Sharon Barnartt, Kay Schriner, and Richard Scotch, "Advocacy and Political Action," in Gary L. Albrecht, K. D. Seelman and M. Bury, eds., *Handbook of Disability Studies* (Oakland, CA: Sage 2001), 430–49.
69. Their protest was sympathetically described by Oliver Sacks in *Seeing Voices* (New York: HarperCollins, 1990).
70. Barnartt, Schriner, and Scotch, "Advocacy," 441.
71. Annetine C. Gelijns, L. D. Brown, C. Magnell, E. Ronchi, and A. J. Moskowitz, "Evidence, Politics, and Technological Change," *Health Affairs* 24 (2005): 29–40.
72. David Sackett, "Evidence-Based Medicine: What It Is and What It Isn't," *British Medical Journal* 312 (1996): 71–72.
73. For social scientists' critique, see Catherine Pope, "Resisting the Evidence: The Study of Evidence-Based Medicine as a Contemporary Social Movement," *Health* 7 (2003): 267–82.
74. On the other hand, guidelines and review articles are also said to be an essential source of confidence for the individual physician, confronted by the unique patient. Though the doctor can never be totally sure how a particular individual will react, he or she has to feel—and convey—a degree of confidence. For this point of view,

see Kathryn Montgomery Hunter, *Doctors' Stories: The Narrative Structure of Medical Knowledge* (Princeton, NJ: Princeton University Press, 1991), 34.

75. Patricia A. Kaufert, "Screening the Body: The Pap Smear and the Mammogram," in Margaret Lock, Allan Young, and Alberto Cambrosio, eds., *Living and Working with the New Medical Technologies* (Cambridge: Cambridge University Press, 2000), 165–83.
76. On this, see Brian Salter and Mavis Jones, "Regulating Human Genetics: The Changing Politics of Biotechnology Governance in the European Union," *Health, Risk, and Society* 4 (2002): 328–40.
77. Daniel Callahan, "Social Sciences and the Tasks of Bioethics," *Daedelus* 128 (1999): 275–94. For the various perspectives on the origins of bioethics, see Renée C. Fox and Judith P. Swazey, *Observing Bioethics* (New York: Oxford University Press, 2008).
78. Fox and Swazey, *Observing Bioethics*, 135–40.
79. Ibid., 170.
80. Callahan, "Social Sciences," 280–81.
81. See Fox and Swazey, *Observing Bioethics*, 215–32.
82. George Weisz, "From Clinical Counting to Evidence-Based Medicine," in G. Jorland, A. Opinel, and G.Weisz, eds., *Body Counts: Medical Quantification in Historical and Sociological Perspectives* (Montreal, Kingston: McGill-Queen's University Press, 2005), 377–93.
83. For the example of regulatory assessment of pharmaceuticals, see John Abraham and Courtney Davis, "Risking Public Safety: Experts, the Medical Profession and 'Acceptable' Drug Injury," *Health, Risk, and Society* 7 (2005): 379–95; Abraham and Davis, "Deficits, Expectations, and Paradigms in British and American Drug Safety Assessments: Prising Open the Black Box of Regulatory Science," *Science, Technology, and Human Values* 32 (2007): 399–431.
84. William Gardner, "Can Human Genetic Enhancement Be Prohibited?" *Journal of Medicine and Philosophy* 20 (1995): 65–84.
85. Mavis Jones and Brian Salter, "The Governance of Human Genetics: Policy Discourse and Constructions of Public Trust," *New Genetics and Society* 22 (2003): 21–41.
86. Jones and Salter, "Governance of Human Genetics," 39.
87. Anne Kerr and Sarah Franklin, "Genetic Ambivalence: Expertise, Uncertainty, and Communication in the Context of New Genetic Technologies," in Webster, *New Technologies*, 40–53.
88. Stefan Timmermans and Marc Berg, "The Practice of Medical Technology," *Sociology of Health and Illness* 25 (2003): 97–114.
89. P. Scott, E. Richards, and B. Martin, "Captives of Controversy: The Myth of the Neutral Social Researcher in Contemporary Scientific Controversies," *Science Technology and Human Values* 15 (1990): 474–94.
90. Harry Collins, "Captives and Victims: Comments on Scott, Richards, and Martin," *Science, Technology, and Human Values* 16 (1991): 249–51.
91. Renée R. Anspach and Nissim Mizrachi, "The Field Worker's Fields: Ethics, Ethnography, and Medical Sociology," *Sociology of Health and Illness* 26 (2006): 713–31.
92. Fox and Swazey, *Spare Parts*, appendix.
93. Paul Rabinow, *Making PCR: A Story of Biotechnology* (Chicago: University of Chicago Press, 1996).
94. Ibid., 20.
95. Renato Rosaldo, *Culture and Truth: The Remaking of Social Analysis* (London: Routledge, 1993), 21.

96. Sue E. Estroff, "Whose Story Is It Anyway?" in S. K. Toombs, D. Barnard, and R. A. Carson, eds., *Chronic Illness: From Experience to Policy* (Bloomington: Indiana University Press, 1995), 77–104.
97. Anspach and Mizrachi, "The Field Worker's Fields," 721–22.
98. Deborah Heath, "Bodies, Antibodies, and Modest Interventions," in Gary L. Downey and Joseph P. Dumit, eds., *Cyborgs and Citadels* (Santa Fe: SAR Press, 1997), 67–82.
99. Gary L. Downey and Joseph P. Dumit, "Locating and Intervening: An Introduction," in Downey and Dumit, *Cyborgs and Citadels*, 5–29.
100. Emily Martin, *Flexible Bodies* (Boston: Beacon Press, 1994); E. Martin, L. Oaks, K.-S. Taussig, and A. van der Straten, "AIDS, Knowledge, and Discrimination in the Inner City," in Downey and Dumit, *Cyborgs and Citadels*, 49–65.
101. Martin et al., "AIDS, Knowledge, and Discrimination," 65.

Chapter 2 **The Making of the Cochlear Implant**

1. See, for example, Hallowell Davis, "The Electrical Phenomena of the Cochlea and the Auditory Nerve," *Journal of the Acoustical Society of America* 6 (1935): 205–15.
2. E. G. Wever and C. W. Bray, "Auditory Nerve Impulses," *Science* 71 (1930): 215.
3. S. S. Stevens, "On Hearing by Electrical Stimulation," *Journal of the Acoustical Society of America* 8 (1937): 191.
4. M. H. Lurie, "Participant Remarks," *Annals of Otology* 81 (1972): 513.
5. A. Djourno and C. Eyries, "Prosthèse auditive par excitation électrique à distance du nerf sensorial à l'aide d'un bobinage inclus à demeure," *Presse Médicale* 65 (1957): 63.
6. William F. House, "Discussant Remarks," *Annals of Otology* 82 (1973): 516; House, "A Personal Perspective on Cochlear Implants," in R. A. Schindler and M. M. Merzenich, eds., *Cochlear Implants* (New York: Raven Press, 1985), 13–15.
7. What follows is based on F. Blair Simmons et al., "Electrical Stimulation of the Acoustic Nerve and Inferior Colliclus in Man," *Archives of Otolaryngology* 79 (1964): 559; Simmons et al., "Auditory Nerve: Electrical Stimulation in Man," *Science* 148 (1965): 104; Simmons, "History of Cochlear Implants in the United States: A Personal Perspective," in Schindler and Merzenich, *Cochlear Implants*, 1; and an interview with Blair Simmons conducted by Jack Spaapen in San Francisco in 1992.
8. Simmons, "History of Cochlear Implants."
9. Kirk Jeffrey, *Machines in Our Hearts: The Cardiac Pacemaker, the Implantable Defibrillator, and American Health Care* (Baltimore: Johns Hopkins University Press, 2001).
10. Fox and Sweezey, *Spare Parts*, chapter 6. Cooley implanted the first artificial heart in April 1969.
11. For an account of Kolff's life and work, see Broers, *Doktor Kolff.*
12. Renée C. Fox and Judith P. Swazey, *The Courage to Fail* (Chicago: University of Chicago Press, 1974), 118.
13. W. F. House and J. Urban, "Long-term Results of Electrode Implantation and Electronic Stimulation of the Cochlea in Man," *Annals of Otology* 82 (1973): 504.
14. M. M. Merzenich and F. A. Sooy, *Report of a Workshop on Cochlear Implants* (San Francisco: University of California, 1974).
15. This account is based on an interview with Merzenich held in Nottingham in September 1992.
16. Interview with Merzenich.
17. William F. House, *Cochlear Implants: My Perspective* (Newport Beach, CA: All-Hear Inc., 1995), 4.

18. Nelson Y. S. Kiang, "Discussant Remarks," *Annals of Otology* 82 (1973): 511–12.
19. W. H. Dobelle, "Discussant Remarks," *Annals of Otology* 82 (1973): 517.
20. Claude-Henri Chouard, *Entendre sans Oreilles* (Paris: Robert Laffont, 1978), 78.
21. June Epstein, *The Story of the Bionic Ear* (Melbourne: Hyland House, 1989), 33.
22. Chouard, *Entendre*, 135.
23. In April 1974, *Le Parisien Libéré* carried a cover story over six columns: "ESPOIR pour 2 million de sourds et 17000 sourds-muets. L'implantation d'un microstimulateur auditief peut leur rendre l'audition et la parole. C'est la révélation faite par les professeurs CHOUARD, PIALOUX et MACLEOD" (Hope for 2 million deaf and 17,000 deaf mutes. The implantation of an auditive microstimulator can give them hearing and speech. That is the revelation made by Professors Chouard, Pialoux, and MacLeod).
24. *L'Express* (May 27, 1974): "Victoire sur la surdité totale. Deux chercheurs francais trouvent le moyen de guérir les sourds-muets de naissance . . . La surdité totale est vaincue . . . Depuis un an, en effet, quinze ex-sourds profonds parlent, entendent. Apres deux mois de reeducation" (Victory over total deafness. Two French researchers find the means of curing people born as deaf-mutes. Total deafness has been vanquished . . . After one year, in effect, 15 ex-deaf are talking, hearing. After two months of rehabilitation).
25. D. Albinhac, *Les Implants Cochleaires: Contribution à l' Histoire de l'Experimentation Humaine* (thesis, Ecole Nationale de la Santé Publique, Rennes, 1978).
26. J. C. Ballantyne, E. F. Evans, and A. W. Morrison, "Electrical Auditory Stimulation in the Management of Profound Hearing Loss," *Supplement to Journal of Laryngology and Otology* (1978).
27. A study carried out in 1978 quotes the head of the ENT clinic at the Paris school for the deaf (INJS): "No 'big name' in French otolaryngology supports Professor Chouard and his group," quoted by Albinhac, *Les implants*, 103.
28. Interview with Chouard, Paris, June 1992.
29. Interview with Douek, London, May 1992.
30. A.R.D. Thornton, ed., *A Review of Artificial Auditory Stimulation* (Southampton: Institute of Sound and Vibration Research, 1977).
31. Graeme Clark, *Sounds from Silence: Graeme Clark and the Bionic Ear Story* (St. Leonards, NSW: Allen and Unwin, 2000).
32. Ibid., 144.
33. Ibid., 52.
34. Metcalfe and Pickstone, "Replacing Hips," 155.
35. R. C. Bilger et al., "Evaluation of Patients Presently Fitted with Implanted Auditory Prostheses," *Annals of Otology, Rhinology, and Laryngology* 86 suppl. 38 (1977): 92–140.
36. "There seems to have been inadequate appreciation and application of the basic physiological and psychophysical information already available, on the processing of speech sounds at peripheral levels of the auditory system, by many of those controlling the implant programs. This has meant that the expected information-carrying capabilities of single channel stimulation (e.g., prosodic and voicing cues) have been but little exploited . . . It is surprising and extremely disappointing to note that unequivocal quantitative data on the benefits of cochlear implantation [compared to high-powered hearing aids and vibrotactile aids] are as yet not available." Ballantyne, Evans, and Morrison, "Electrical Auditory Stimulation," 91.
37. R. Garud and A. H. Van de Ven, "Technological Innovation and Industry Emergence: The Case of Cochlear Implants," in A. H. Van de Ven, H. L. Angle, and M. S. Poole, eds., *Research on the Management of Innovation: The Minnesota Studies* (New York: Harper and Row, 1989), 503.

38. This discussion of 3M's cochlear implant activities is based on A. H. Van de Ven, D. E. Polley, R. Garud, and S. Venkataraman, *The Innovation Journey* (New York: Oxford University Press, 1999).
39. Ibid., 227.
40. Ibid., 236.
41. B. J. Edgerton, W. F. House, J. A. Brimacombe, and L. S. Eisenberg, "Status of the Cochlear Implant Program at the House Ear Institute," *Advances in Audiology* 2 (1984): 68–89.
42. "Nucleus," Douek told me, "sent someone to work in Fourcin's laboratory and to understand what he was trying to do. And the concepts were copied wholesale into their equipment. It worked. It was not things you could patent, and we weren't thinking of patenting . . . and also, this is England. And initially when they read papers they always gave credit . . . Gradually the credit became less and less" (Douek interview). There is no mention of this in Clark's book. He writes only of having decided *not* to spend a sabbatical leave in Fourcin's lab because of IRA bombings in London at the time. Instead, he went to Keele University.
43. Crucial here was a meeting with Ernst Lehnhardt, an ear surgeon from Hannover in Germany, who visited Melbourne early in 1984. Within a couple of years, Lehnhardt had created the largest implant center in Europe. In 1987, Dr. Monika Lehnhardt became chief executive of the manufacturer's new European subsidiary, Cochlear AG. See Clark, *Sounds from Silence*, 162.
44. The Hochmairs themselves were at the same time developing multichannel intracochlear devices and would soon afterward establish their own company, MedEl, to commercialize their designs.
45. Van de Ven et al., *Innovation Journey*, 262.
46. The outside processor weighed 600 grams and measured 16 x 15 centimeters.
47. Interview with Douek. By the late 1980s, Bertin had decided to sell its cochlear implant technology and in 1988 did so.
48. Interview with Douek.
49. See Susan Bartlett Foote, *Managing the Medical Arms Race: Innovation and Public Policy in the Medical Device Industry* (Berkeley: University of California Press, 1992), 162.
50. The discussion that follows is based on N. M. Kane and P. D. Manoukian, "The Effect of the Medicare Prospective Payment System on the Adoption of New Technology: the Case of Cochlear Implants," *New England Journal of Medicine* 321(1989): 1378–83.
51. Clark, *Sounds from Silence*, 160. Soon after this, the company took over the implant interests of 3M and Symbion (which had developed a device known as the Ineraid, based on work done at the University of Utah and later marketed by Johnson and Johnson).
52. Garud and Van de Ven, "Technological Innovation," 504.
53. A.R.D. Thornton, "Estimation of the Number of Patients Who Might Be Suitable for Cochlear Implant and Similar Procedures," *British Journal of Audiology* 20 (1986): 221–29. Many people have hearing losses of between ten and thirty decibels. Hearing loss of up to seventy decibels can typically be helped with a hearing aid. "Deafness" is commonly defined as a hearing loss of approximately one hundred decibels within the range of frequencies used in human speech. Later Davis, Fortnum, and O'Donoghue estimated that in Europe, 2,268 children per annum could be candidates for an implant. This implied a Europe-wide expenditure of some £17 million per annum. See A. Davis, H. Fortnum, and G. O'Donoghue, "Children Who Could Benefit from a Cochlear Implant: A European Estimate of

Projected Numbers, Cost, and Relevant Characteristics," *International Journal of Pediatric Otorhinolaryngology* 31 (1995): 221–33.

54. Gerald E. Loeb, "Remarks," in W. F. House, L. Grammatico, C. Fugain, G. E. Loeb, E. E. Douek, and F. Blair Simmons, "Cochlear Implants in Children: Panel Discussion," in Schindler and Merzenich, *Cochlear Implants*, 575–87.
55. National Institutes of Health, "Consensus Development Conference Statement on Cochlear Implants," *Archives of Otolaryngology Head and Neck Surgery* 115 (1989): 31–36.
56. E. Owens and D. Kessler, eds., *Cochlear Implants in Young Deaf Children* (Boston: Little, Brown, 1989).
57. Food and Drug Administration, *Summary of Safety and Effectiveness Data, Premarket Approval Application P890027*, June 27, 1990.
58. National Institutes of Health, *Cochlear Implants in Adults and Children*, NIH Consensus Statement, vol. 13, no. 2, 1995.
59. In the United States, approximately 23,000 adults and 15,500 children had been implanted by the end of 2006 (http://www.nidcd.nih.gov/health/hearing/coch.asp, accessed August 2008).
60. In 1988, "serial entrepreneur" Alfred E. Mann had entered into a license agreement with UCSF for the right to make, use, and sell the inventions the university had developed over the previous fifteen years. Intrigued by the notion of "making deaf people hear," the cochlear implant was selected as the first invention to be brought to market. This was to become the Clarion implant, and a spin-off firm, MiniMed, was established to produce and market it (as well as insulin pumps and glucose sensors). In 1993, Mann merged MiniMed Technologies and other of his activities into a new company called Advanced Bionics. In March 1996, this company received approval from the FDA to market the Clarion for use in postlingually deafened adults and in 1997, for use in children. In 2004, Advanced Bionics was taken over by Boston Scientific, a major international producer of scientific and medical instruments. See M. J. Coyle, "Confessions of a Serial Entrepreneur: A Conversation with Alfred E. Mann," *Health Affairs* 25 (2006): 104–13.
61. M. V. Goycoolea and the Latin American Cochlear Implant Group, "Latin American Experience with the Cochlear Implant," *Acta Oto-Laryngologica* 125 (2005): 468–73. I am grateful to Beatriz Miranda for bringing this article to my attention.
62. In late 2005, an Indian national newspaper, *The Hindu*, carried an article with the title "Indigenous Cochlear Implant to Hit the Market Soon." Developed by the Defense Research and Development Organization and local medical professionals, the Indian implant was expected to cost only one-fifth of the cheapest device available. B. S. Perappadan, "Indigenous Cochlear Implant," http://www.thehindu.com/2005/11/25/stories/ (consulted September 2008). Goycoolea notes that experimental prototypes were developed in Latin America as early as 1975, though "these were eventually replaced by established commercial implants." Gooycolea, "Latin American Experience,"471. According to Zeng, China and Korea have working prototypes: Fan-Gang Zeng, "Viewpoint," *Hearing Journal* 60 (2007): 48–49.
63. Coyle, "Confessions."

Chapter 3 **The Cochlear Implant and the Deaf Community**

1. Sherry Adler found that of breast cancer sufferers in the Bay Area, 69 percent had used CAM before their diagnosis, compared with 72 percent afterward. See Sherry R. Adler, "Complementary and Alternative Medicine Use among Women with Breast Cancer," *Medical Anthropology Quarterly* 13 (1999): 214–22.

2. $27 billion in 1997. See D. M. Eisenberg, R. B. Davis, S. L. Ettner, et al., "Trends in Alternative Medicine Use in the United States, 1990–1997," *Journal of the American Medical Association* 280 (1998): 1569–75.
3. Phil Brown, S. Zavestoski, S. McCormick, B. Mayer, R. Morello-Frosch, and R. Gasior Altman, "Embodied Health Movements: New Approaches to Social Movements in Health," *Sociology of Health and Illness* 26 (2004): 50–80.
4. What follows is taken from the Wikipedia entry on pro-ana, http://en.wikipedia.org/wiki (consulted October 21, 2008).
5. A sociological study of one such pro-ana community ("Anagrrl") concluded that many of the girls and young women "cling to anorexia as a 'macabre comfort'; many do not want to eradicate the illness, but want to find safe ways to live with it." For many of them, living with anorexia provided a minimal kind of security, something to hold on to, however dangerous. Katie Ward, Mark Davis, and Paul Flowers, "Patient 'Expertise' and Innovative Health Technologies," in Webster, *New Technologies*, 97–111.
6. Alex Broom and Philip Tovey, "Therapeutic Pluralism? Evidence, Power, and Legitimacy in UK Cancer Services," *Sociology of Health and Illness* 29 (2007): 551–69.
7. Nissim Mizrachi, Judith T. Shuval, and Sky Gross, "Boundary at Work: Alternative Medicine in Biomedical Settings," *Sociology of Health and Illness* 27 (2005): 20–43; Broom and Tovey, "Therapeutic Pluralism."
8. Wikipedia, http://en.wikipedia.org/wiki/Pro-ana (consulted October 21, 2008).
9. Fan quoted in Robert M. Sade, "Complementary and Alternative Medicine: Foundations, Ethics, and Law," *Journal of Law, Medicine, and Ethics* 31 (2003): 183–90.
10. Paul C. Higgins, *Outsiders in a Hearing World: A Sociology of Deafness* (Thousand Oaks: Sage Books, 1980), 14.
11. Ibid., 68.
12. Some authors use "Deaf" to refer to the culturally deaf and "deaf" to refer to the audiologically deaf. There are many people who are audiologically deaf, for example as a result of aging, but who neither sign nor identify with the Deaf community. Here I shall use "Deaf" only when referring to the Deaf community.
13. Harlan Lane, *When the Mind Hears* (New York: Random House, 1984). See also Jonathan Rée, *I See a Voice: A Philosophical History* (London: Flamingo, 1999).
14. See Douglas C. Baynton, *Forbidden Signs: American Culture and the Campaign against Sign Language* (Chicago: Chicago University Press, 1996).
15. M. M. Maxwell and P. Kraemer, "Speech and Identity in the Deaf Narrative," *Text* 10 (1990): 339.
16. William C. Stokoe, afterword to Charlotte Baker and Robin Battison, eds., *Sign Language and the Deaf Community: Essays in Honor of William C. Stokoe* (Silver Springs, MD: National Association of the Deaf, 1980), 265.
17. Ibid., 266–67.
18. Edward Klima and Ursula Bellugi, *The Signs of Language* (Cambridge, MA: Harvard University Press, 1979); Howard Poizner, Edward Klima, and Ursula Bellugi, *What the Hands Reveal about the Brain* (Cambridge, MA: MIT Press, 1987).
19. John C. Marshall, introduction to Poizner, Klima, and Bellugi, *What the Hands Reveal*, xiii.
20. Bernard Mottez and Harry Markowicz, "Integration ou droit à la difference: Les conséquences d'un choix politique sur la structuration et le mode d'existence d'un groupe minoritaire, les sourds" (unpublished report, Paris: CEMS, 1979).

21. Bernard T. Tervoort, "Gebaren en gebarentaal in Europa," in B. T. Tervoort, ed., *Hand over Hand: Nieuwe Inzichten in de Communicatie van Doven* (Muiderberg: Dick Coutinho, 1983), 55–69.
22. Barbara Kannapell in Baker and Battison, *Sign Language*, 109.
23. Carol Padden and Tom Humphries, *Deaf in America: Voices from a Culture* (Cambridge, MA: Harvard University Press, 1988).
24. Ibid., 29.
25. "For the last thirty-three years, the French Deaf people have watched with tear-filled eyes and broken hearts this beautiful language of signs snatched away from their schools. For the last thirty-three years they have striven and sought to reinstate signs in the schools, but for thirty-three years their teachers have cast them aside and refused to listen to their pleas." Ibid., 34–36.
26. This is not everywhere the case. In Germany, as Linda Hogle's comparative analysis makes very clear, the instrumentalization of death and the risks to self-determination—in other words, the costs of organ transplantation—received much more media attention. See Linda F. Hogle, *Recovering the Nation's Body: Cultural Memory, Medicine, and the Politics of Redemption* (New Brunswick: Rutgers University Press, 1999). In Japan, attitudes to transplantation diverge still further from those in the United States. See, for example, Margaret Lock, *Twice Dead: Organ Transplants and the Reinvention of Death* (Berkeley: University of California Press, 2001).
27. Rebecca Dresser, *When Science Offers Salvation: Patient Advocacy and Research Ethics* (New York: Oxford University Press, 2001), 131.
28. Ibid., 135.
29. Chouard, *Entendre*, 124–25.
30. Administrative Council of UNISDA, document dated April 16, 1977, quoted by Albinhac, *Les Implants*, 94–95.
31. M. F. "L'oreille artificielle est prématurée et dangereuse," *L'Aurore*, February 27, 1979.
32. G. Jones, "Electrodes Work for Deaf Girl," *Daily Telegraph*, August 16, 1984.
33. G. Jones, "Hope of Restoring Deaf Girl's Hearing with Electrodes Implant," *Daily Telegraph*, August 13, 1984.
34. S. O'Hagan, "Cochlear Implants: A Cause for Concern," *Talk* (Autumn 1984): 10–11.
35. "Surgeons Warned on Ear Operation," *The Standard*, October 22, 1984.
36. Carlo Laurenzi, "The Bionic Ear and the Mythology of Paediatric Implants," *British Journal of Audiology* 27 (1993): 1–5.
37. Clark, *Sounds from Silence*, 71.
38. Intriguingly, critics made little of the futuristic associations of the term. An exception was the respected British psychologist George Montgomery. In a somewhat tongue-in-cheek address to the British Deaf Association, Montgomery referred to the possibility of external control of the information received by an implantee: "Implants within the human brain itself have long been in use but have received more popular attention recently with the proposal to implant a microchip in the visual cortex of airline pilots to provide 'head up' displays of flight information and intelligence up-dated to the micro-second from ground control. No reports have been received yet for a similar 'head up' display of the 10 commandments to be implanted in criminals . . . With the criminally insane, however, destructive and self-destructive, violence has long justified the implant of radio-controlled induction coils in the brain areas in the allocortex, transitional cortex and tip of the temporal lobe which inhibit or restrain hyothalamic and sub-cortical centers of emotion giving rise to violent behavior." See George Montgomery, "Bionic

Miracle or Megabuck Acupuncture? The Need for a Broader Context in the Evaluation of Cochlear Implants," in M. D. Garretson, ed., *Perspectives in Deafness: A Deaf American Monograph* (Silver Springs, MD: National Association of the Deaf, 1991), 97–105.

39. Rebecca Dresser provides comparable examples of health advocacy groups responding to exaggerated accounts of "breakthroughs" in cancer treatment. See Dresser, *Science Offers Salvation*, 132–33.
40. Josette Chalude, "Deaf Power," *Communiquer* 62 (1982): 3–4.
41. Quoted in Albinhac, *Les Implants*, 86. See also Jean Grémion, *La Planète des Sourds* (Paris: Sylvie Messinger, 1990).
42. "Signed English" is English expressed in signs. Sign languages have a different grammatical structure from spoken languages. Signed English uses the grammar of the spoken rather than the sign language.
43. Heather Mohay, "Deafness in Children," *Medical Journal of Australia* 154 (1991): 372–74.
44. A. J. Kelly, letter to the editor, *Medical Journal of Australia* 154 (1991): 491.
45. W.P.R. Gibson, "Opposition from Deaf Groups to the Cochlear Implant," *Medical Journal of Australia* 155 (1991): 212–14.
46. Harlan Lane, *The Mask of Benevolence: Disabling the Deaf Community* (New York: Alfred Knopf, 1992).
47. Theodore M. Porter, *Trust in Numbers: The Pursuit of Objectivity in Science and Public Life* (Princeton, NJ: Princeton University Press, 1995).
48. Desmond J. Power and Mervyn P. Hyde, "The Cochlear Implant and the Deaf Community," *Medical Journal of Australia* 157 (1992): 421–22.
49. The National Union of the Deaf was a small radical organization active in the 1970s and 1980s. See Paddy Ladd, "Oralism's 'Final Solution,'" *British Deaf News* (Autumn 1985): 5–6.
50. E. Dolnick, "Deafness as Culture," *Atlantic Monthly*, September 1993, 43.
51. Lane, *Mask of Benevolence.*
52. For a more recent expression of Harlan Lane's views, see Harlan Lane, "Ethnicity, Ethics, and the Deaf-World," *Journal of Deaf Studies and Deaf Education* 10 (2005): 291–310.
53. Harlan Lane and Benjamin Bahan, "Ethics of Cochlear Implantation in Young Children: A Review and Reply from a Deaf-World Perspective," *Otolaryngology Head and Neck Surgery* 119 (1998): 297–313.
54. Robert A. Crouch, "Letting the Deaf be Deaf: Reconsidering the Use of Cochlear Implants in Prelingually Deaf Children," *Hastings Center Report* 27 (1997): 14–21.
55. Lane and Bahan, "Ethics."
56. It is then comparable with the issues raised by transcultural adoption, a line of argument that they do not pursue.
57. N. L. Cohen, "Cochlear Implants in Young Children: Ethical Considerations," *Annals of Otology, Rhinology, and Laryngology* 104 suppl. 166 (1995): 17–19.
58. Lane and Bahan, "Ethics."
59. For example, it was of importance to patient advocates demanding wider criteria of access to clinical trials. See Dresser, *Science Offers Salvation*, 52–62.
60. Nora Jacobson, *Cleavage: Technology, Controversy, and the Ironies of the Man-made Breast* (New Brunswick: Rutgers University Press, 2000), 195.
61. National Association of the Deaf, *Cochlear Implants in Children* (Silver Springs, MD: National Association of the Deaf, 1991).
62. Alonso Plough, *Borrowed Time: Artificial Organs and the Politics of Extending Lives* (Philadelphia: Temple University Press, 1986), 19.

63. Lane, "Ethnicity."
64. Quoted by Dolnick, "Deafness as Culture," 43.

Chapter 4 **The Globalization of a Controversial Technology**

1. Cochlear Corporation, *Annual Report for 2007* (http://www.cochlea.com).
2. Benedicte Ingstad and Susan Reynolds Whyte, eds., *Disability in Local and Global Worlds* (Berkeley: University of California Press, 2007), 10.
3. J. Rogers Hollingsworth, J. Hage, and R. A. Hanneman, *State Intervention in Medical Care: Consequences for Britain, France, Sweden, and the United States, 1890–1970* (Ithaca: Cornell University Press, 1990).
4. Parthasarathy, *Building Genetic Medicine.*
5. Arjun Appadurai, "Grassroots Globalization and the Research Imagination," in A. Appadurai, ed., *Globalization* (Durham, NC: Duke University Press, 2001), 5–6.
6. In addition to 3M/House and MedEl implants, some used the Ineraid device, based on work initially done at the University of Utah and subsequently produced by Johnson and Johnson. It was later withdrawn from the market.
7. Interview with Fraser, London, February 1992.
8. Interview with Fraser.
9. This device was later marketed by the Austrian company MedEl.
10. The RNID at that time was an institution providing services *to* deaf people. It was not an advocacy group established *by* the Deaf community, like the British Deaf Association. It was run (at that time) by doctors and service providers. For a disillusioned insider's view of the objectives and the politics of British organizations of/for the deaf, see Doug Alker, *Really Not Interested in the Deaf* (published by the author, 2000).
11. Mark P. Haggard, "Straight Thinking about Cochlear Implants," *British Journal of Audiology* 20 (1986): 5–7.
12. Interview with Douek.
13. Interview with Fraser.
14. Letter from Harry Cayton, director of the NDCS, to G. M. O'Donoghue FRCS, dated March 2, 1989.
15. British Deaf Association, *Policy on Cochlear Implants* (Carlisle: BDA, 1994).
16. Q. A. Summerfield and D. H. Marshall, *Cochlear Implantation in the UK, 1990–1994,* report by the MRC Institute of Hearing Research, Nottingham (London: Her Majesty's Stationery Office, 1995).
17. Ibid., 155.
18. Ibid., 261.
19. "That [FDA approval] was very important to us. Because, well, we had to implant the apparatus, to import it. We said to ourselves 'what if problems arise, infection, leakage?' If that happens it is extremely advantageous to be able to refer to American experience, to the FDA approval. We had the feeling that we would then be adequately covered if anything went wrong" (interview with Huizing, Utrecht, 1992).
20. For example, there was an article in a popular women's magazine *Margriet* with the title "I Can Hear Again," which describes her parents' satisfaction with the progress being made by five-year-old Birgitta.
21. Jan Brouwer de Koning, "Elektrisch binnenoor overtreft gebarentaal," *Trefpunt* 8 (1993): 16–17.
22. Sam Pattipeiluhu, "Cochleaire implantatie: De ondergang van Dovencultuur?" *Woord en Gebaar* 15 (1995): 15–16.

23. Academisch Ziekenhuis Nijmegen, *Cochleaire implantatie bij kinderen van mart 1993 tot mart 1996*. Eindverslag van het ontwikkelingsgeneeskunde project (Nijmegen/St-Michielsgestel/Amsterdam, February 1996), 124.
24. Interview with Bredberg, Stockholm, March 1992.
25. Cued speech is a system that uses specially developed signs to distinguish speech sounds that cannot be visually distinguished. Although signs are used, it bears no relation to sign language.
26. I am grateful to Gunilla Öhngren for her help in collecting Swedish material.
27. Interview with Risberg, Stockholm, March 1992.
28. These included, beyond Nucleus and MedEl devices, the Laura, which had been developed by Belgian researchers. It was produced briefly by Philips but was later withdrawn from the market.
29. European Union of the Deaf (EUD) Working Group on Cochlear Implants, *Final Report* (Brussels: European Union of the Deaf, 1996).
30. Very little research has been done on attitudes to deafness in the developing world, but see D. Stephens, R. Stephens, and A. von Eisenhart-Rothe, "Attitudes toward Hearing-Impaired Children in Less Developed Countries: A Pilot Study," *Audiology* 39 (2000): 184–91.
31. Peter Conrad and Valerie Leiter, "Medicalization, Markets, and Consumers," *Journal of Health and Social Behavior* 45 special issue (2004): 158–76.
32. This material on the Czech Republic was collected by František Bouška, under the guidance of Karel Muller of the Charles University. I am grateful to both for their help.
33. Fan-Gang Zeng, "Cochlear Implants in Developing Countries," *Contact* 10 (1996): 5–9.
34. M.I.J. Khan, N. Mukhtar, S. R. Saeed, and R. T. Ramsden, "The Pakistan (Lahore) Cochlear Implant Programme: Issues Relating to Implantation in a Developing Country," *Journal of Laryngology and Otology* 121 (2007): 745–50.
35. The New Vision, http://www.newvision.co.ug/D/8/12/613955.
36. The Deaf-L list was established by Roy Miller, a political scientist from Southern Illinois University, who became deaf in 1987 at the age of forty-seven. It was run by him until his retirement in 1999. The World Federation of the Deaf was established in 1951 as an association of national federations. It holds conferences every four years and is currently headquartered in Helsinki, Finland. See http://en.wikipedia.org/wiki/World_Federation_of_the_Deaf.
37. Fox and Swazey, *Observing Bioethics*, 217.
38. After its takeover, Advanced Bionics became the neuromodulation division of Boston Scientific. See http://www.cnw.ca/fr/releases/archive/June2007/21/c7593.html (consulted September 2008).
39. Cochlear Corporation, *Annual Report* for 2007.
40. For example, in the 1960s antinuclear groups set about "reframing" nuclear power as a source of unacceptable risk rather than of cheap energy. Examining shifts in public opinion regarding nuclear power in different countries and changes in national energy policies, Koopmans and Duyvendak concluded that relative success and failure had to be understood in terms of political processes and power relations. See Ruud Koopmans and Jan Willem Duyvendak, "The Political Construction of the Nuclear Energy Issue and Its Impact on the Mobilization of Antinuclear Movements in Western Europe," *Social Problems* 42 (1995): 235–51.
41. In Britain there had been the National Union of the Deaf in the 1970s and 1980s, now seen as having "opened the door" for its radical successors including (today) the Federation of Deaf People and the Deaf Liberation Front (both founded at the end of the 1990s).

42. Sourds en Colère, *Dossier* (Paris, undated).
43. See Laborit's autobiography, Emmanuelle Laborit, *The Cry of the Gull* (Washington, DC: Gallaudet University Press, 1998).
44. For an account of the establishment and procedures of the French national ethics committee, see Fox and Swazey, *Observing Bioethics*, 233–58.
45. Laurence Folléa, "Querelle de langage chez les sourds," *Le Monde*, May 26, 1994.
46. Comité Consultatif National d'Ethique pour les Sciences de la Vie et de la Santé (CCNE), *Avis sur l'implant cochleaire chez l'enfant sourd pré-lingual* (Paris: CCNE, December 1994), 44.
47. Dominique Schnapper, *La France de l'Intégration: Sociologie de la Nation en 1990* (Paris: Editions Gallimard, 1991).
48. Hilde Haualand, "The Two-Week Village: The Significance of Sacred Occasions for the Deaf Community," in Ingstand and Whyte, *Disability*, 33–55.
49. Khan et al., "Pakistan Cochlear Implant Programme."
50. Zeng, "Cochlear Implants in Developing Countries."
51. Leila Monaghan, Constanze Schmaling, Karen Nakamura, and Graham H. Turner, eds., *Many Ways to Be Deaf: International Variation in Deaf Communities* (Washington, DC: Gallaudet University Press, 2003).
52. Karen Nakamura, "U-turns, Deaf Shock, and the Hard of Hearing: Japanese Deaf Identities at the Borderlands," in Monaghan, *Many Ways to Be Deaf*, 211–29; Karen Nakamura, *Deaf in Japan: Signing and the Politics of Identity* (Ithaca: Cornell University Press, 2006).
53. James Woodward, "Sign Languages and Deaf Identities in Thailand and Viet Nam," in Monaghan, *Many Ways to Be Deaf*, 283–301.
54. Richard J. Senghas, "New Ways to Be Deaf in Nicaragua: Changes in Language, Personhood, and Community," in Monaghan, *Many Ways to Be Deaf*, 260–82; Laura Polich, *The Emergence of the Deaf Community in Nicaragua* (Washington, DC: Gallaudet University Press, 2005).

Chapter 5 **Implantation Politics in the Netherlands**

1. L.B.W. Jongkees, "Doven weer horen?" *Nederlands Tijdschrift voor Geneeskunde* 122 (1978): 1621.
2. L.B.W. Jongkees, "Doven weer horen?" *Nederlands Tijdschrift voor Geneeskunde* 126 (1982): 1459.
3. Interview with Huizing.
4. Interview with Huizing.
5. E. H. Huizing and G. F. Smoorenburg, eds., *De Elektrische Binnenoorprosthese* (Utrecht: Nederlands Vereniging voor Audiologie, 1986).
6. Academisch Ziekenhuis Nijmegen/Instituut voor Doven Sint-Michielsgestel, *Cochlear Implants* (Eindverslag project Ontwikkelingsgeneeskunde 1988–1990. Publikatienummer IvD/R&D/9104/01. August 1991), 57.
7. Nijmegen/IvD, *Cochlear Implants*.
8. The approach that I gradually adopted, in interaction with events rather than on the basis of any a priori methodological choice, was an ethnographic one. Characteristic of ethnography is the immersion and participation in the social world being studied that it requires of the researcher. Martyn Hammersley and Paul Atkinson, in their well-known introduction to ethnography, suggest that influence on the social phenomena being studied is inevitable and, differently from other social science methodologies, the ethnographer does not try to minimize these effects but rather to reflect on and understand them. See Martyn Hammersley and Paul Atkinson, *Ethnography: Principles in Practice* (London:

Routledge, 1989). But see Anspach and Mizrachi, "The Field Worker's Fields," for a different perspective on the ethnographer's influence on the phenomena being studied.

9. Jennifer Harris, "Boiled Eggs and Baked Beans: A Personal Account of a Hearing Researcher's Journey through Deaf Culture," *Disability and Society* 10 (1995): 295–308; Higgins, *Outsiders*.
10. The organization is now called Dovenschap.
11. Johan Ros, "C.I. onderzoek, wat is dat?" *Woord en Gebaar* 13 (1993): 18–19.
12. S. Kamerling, "De cochleaire implant: Een hoortoestel waar *niemand* echt horend van wordt," *Doven en Welzijn*, December 15–17, 1993.
13. P. Messer and T. van Corven, "Artsen en doven strijden over 'nieuw paar oren,'" *Trouw*, November 1, 1993.
14. Kerstin Heiling, *The Development of Deaf Children: Academic Achievement Levels and Social Processes* (Hamburg: Signum, 1995).
15. Ibid.
16. The classic study of Dutch political culture is Arend Lijphart, *The Politics of Accommodation: Pluralism and Democracy in the Netherlands* (Berkeley: University of California Press, 1968).
17. Downey and Dumit, *Cyborgs and Citadels*, 16.
18. Discussing her earlier fieldwork in a neonatal intensive care unit, Renée Anspach refers to a tension between what she saw as the dictates of her method and the commitment she felt. "In acting in what I saw as the parents' best interest, I was also consciously violating my field's standards of good ethnographic work. At the same time, I was responding as a human being to a straightforward request for help from another." See Anspach and Mizrachi, "The Field Worker's Fields," 721.
19. Jan Pieter Govers, "Nieuwe oren van de keizer: Realiteit of illusie?" (Ede: unpublished thesis, 1994), 134.
20. This report is discussed in chapter 4; see chapter 4, note 46.
21. Stuart Blume, "Veel gevolgen van cochleaire implantatie zijn nog volstrekt onduidelijk," *Woord en Gebaar* 15 (January 1995): 9–10.
22. Stuart Blume, "Implantatie teams zijn bereid dialoog aan te gaan met de Dovengemeenschap," *Woord en Gebaar* 15 (March 1995): 6.
23. Lianne Vermeulen, Jan Brokx, Anneke Vermeulen, and Paul van den Broek, "Pedagogische aspekten rondom cochleaire implantatie," talk presented at a meeting of the Dutch Association for Philosophy and Medicine, Amsterdam, 1995.
24. Academisch Ziekenhuis Nijmegen, *Cochleaire implantatie bij kinderen van mart 1993 tot mart 1996*. Eindverslag van het ontwikkelingsgeneeskunde project (Nijmegen/St-Michielsgestel/Amsterdam, February 1996).
25. That is to say, the child used signs to clarify its spoken words, following the grammar of the spoken not the signed language.
26. Academisch Ziekenhuis Nijmegen, *Cochleaire implantatie bij kinderen van mart 1993 tot mart 1996*, 124.
27. Maarten Evenblij, "Een beetje doof met een elektrisch oor," *De Volkskrant*, February 3, 1996.
28. Ziekenfondsraad (Health Insurance Council of the Netherlands), *Rapport Cochleaire Implantatie bij Kinderen* (Amstelveen, December 1996).
29. Commissie Nederlands Gebarentaal, *Méér dan een Gebaar*, Rapport van de Commissie Nederlandse Gebarentaal (Den Haag: SDU Uitgevers, 1997).
30. The Council of Europe should not be confused with the European Union. Founded in 1949, the Council of Europe seeks to develop throughout Europe common and democratic principles based on the European Convention on Human Rights and other reference texts on the protection of individuals. See http://www.coe.int.

31. *Sound and Fury*, a documentary made by Josh Aronson and first shown in 2000, was nominated for a "Best Documentary" Academy Award in 2001 and won numerous prizes at film festivals. It portrays the dilemma involved in choosing a cochlear implant.
32. Letter from the author to the Ministry of Health, October 1999.
33. The Health Council (Gezondheidsraad) advises on the state of science in a given field of medicine or health. The Health Insurance Council (Ziekenfondsraad) at that time advised on matters of reimbursement (but has since been replaced by a different body).
34. This report was later published. Gezondheidsraad (Health Council of the Netherlands), *Cochleaire Implantatie bij Kinderen* (Den Haag: Gezondheidsraad Report 21, 2001).
35. Gardner, "Human Genetic Enhancement."
36. Salter and Jones reach a similar conclusion: "If bioethics is to influence the political culture of regulation against the automatic acceptance of the utilitarian ethic of economic advance, then it requires legal and institutional platforms for its activities. But in order to achieve this transition, international bioethics has to have allies at the regional and national levels." Salter and Jones, "Regulating Human Genetics," 329.

Chapter 6 ***Contexts of Uncertainty: Parental Decision Making***

1. See her autobiography, Bonnie Poitras Tucker, *The Feel of Silence* (Philadelphia: Temple University Press, 1995).
2. Bonnie Poitras Tucker, "Deaf Culture, Cochlear Implants, and Elective Disability," *Hastings Center Report* 28, no. 4 (1998): 6–14.
3. D.M.R. Kelsay and R. S. Tyler, "Advantages and Disadvantages Expected and Realized by Pediatric Cochlear Implant Recipients as Reported by Their Parents," *American Journal of Otology* 17 (1996): 866–73.
4. T. H. Sach and D. K. Whynes, "Paediatric Cochlear Implantation: The Views of Parents," *International Journal of Audiology* 44 (2005): 400–407.
5. T. N. Kluwin and D. A. Stewart, "Cochlear Implants for Younger Children: A Preliminary Description of the Parental Decision Process and Outcomes," *American Annals of the Deaf* 145 (2000): 26–32.
6. A. Steinberg, A. Brainsky, L. Bain, L. Montoya, M. Indenbaum, and W. Potsic, "Parental Values in the Decision about Cochlear Implantation," *International Journal of Pediatric Otolaryngology* 55 (2000): 99–107.
7. S. Okubo, M. Takahashi, and I. Kai, "How Japanese Parents of Deaf Children Arrive at Decisions Regarding Cochlear Implantation Surgery: A Qualitative Study," *Social Science and Medicine* 66 (2008): 2436–47.
8. Sach and Whyne, "Paediatric Cochlear Implantation," 405.
9. Okubo, Takahashi, and Kai, "Japanese Parents," 2442.
10. Robert Q. Pollard Jr., "Conceptualizing and Conducting Preoperative Psychological Assessments of Cochlear Implant Candidates," *Journal of Deaf Studies and Deaf Education* 1 (1996): 16–28.
11. Pollard, "Conceptualizing and Conducting," 22.
12. O. Corrigan, "Empty Ethics: The Problem with Informed Consent," *Sociology of Health and Illness* 25 (2003): 768–92.
13. T. Balkany, A. V. Hodges, and K. Goodman, "Additional Comments" following Lane and Bahan, "Ethics."
14. This quotation, and those following, come from interviews conducted in southern England in 1996 in collaboration with Professor Lucy Yardley.

15. B. Freedman, A. Fuks, and C. Weijer, "In Loco Parentis: Minimal Risk as an Ethical Threshold for Research on Children," *Hastings Center Report* 23 (1993): 13–19.
16. Priscilla Alderson has found that children from about eight years of age do commonly wish to share in decision making regarding their treatment and, in many cases (depending on their prior experience), are competent to do so. She argues that their competences are far greater than has generally been assumed. See P. Alderson, "Competent Children? Minors' Consent to Health Care Treatment and Research," *Social Science and Medicine* 65 (2007): 2272–83.
17. Lucy Yardley, "The Quest for Natural Communication: Technology, Language, and Deafness," *Health* 1 (1997): 37–55.
18. John B. Christiansen and Irene W. Leigh, *Cochlear Implants in Children: Ethics and Choices* (Washington, DC: Gallaudet University Press, 2002), 84.
19. Ibid., 309.
20. Ibid.
21. The 1991 position paper to which the parents referred in 1999 interviews was replaced by a different one in 2000.
22. Christiansen and Leigh, *Cochlear Implants*, 266.
23. Roslyn Rosen was a vice president for academic affairs at Gallaudet University and an officer of the World Federation of the Deaf.
24. Rayna Rapp and Faye Ginsburg, "Enabling Disability: Rewriting Kinship, Reimagining Citizenship," *Public Culture* 13 (2001): 533–56.
25. The concept had been attacked in the pages of the French parents' association ANPEDA. See chapter 3.
26. See http://www.cochlear.com.
27. Svetlana Kolchick, "With Implant Miss America May Hear Again," *USAToday.com* (http://www.usatoday.com/life2002–08–07-whitestone_x.htm, consulted October 2002).
28. Charles E. Bosk, "Professional Ethicist Available: Logical, Secular, Friendly," *Daedalus* 128 (Fall 1999): 47–68.
29. Okubo, Takahashi, and Kai, "Japanese Parents," 2442.
30. Sach and Whynes, "Paediatric Cochlear Implantation."
31. Beverly Biderman, *Wired for Sound: A Journey into Hearing* (Toronto: Trifolium Books, 1998), 2.
32. Biderman, *Wired for Sound*, 119.
33. The (British) National Cochlear Implant Users Association has as its objectives "to fight for those people who have cochlear implants, to seek as many improvements as we can and also to campaign for those people who want implants and are denied them." See National Cochlear Implant Users Association, *Cochlear Implants: A Collection of Experiences of Users of All Ages* (High Wycombe, UK: NCIUA, 2001).
34. Thomson, in NCIUA, *Cochlear Implants*, 20–21.
35. Introduced by the French Post Office in 1982, Minitel was a telephone line–based textual communication system that can be seen as a precursor of e-mail.
36. A vivid account of the meaning of a bodily mark being transformed as the bearer moves from one culture to another is Aud Talle's account of the experience of circumcised Somali women forced into exile in London. See A. Talle, "From 'Complete' to 'Impaired' Body: Female Circumcision in Somalia and London," in Ingstad and Whyte, *Disability*, 56–77.
37. See Nakamura, *Deaf in Japan*; Jan-Kåre Breivik, *Deaf Identities in the Making* (Washington, DC: Gallaudet University Press, 2005).
38. Christiansen and Leigh, *Cochlear Implants*, 213.

39. Ibid., 321.
40. Okubo, Takahashi, and Kai, "Japanese Parents," 2444.
41. Eric Parens and Adrienne Asch, "The Disability Rights Critique of Prenatal Testing: Reflections and Recommendations," *Hastings Center Report* 29 suppl. (1999): S1-S22.
42. Christiansen and Leigh, *Cochlear Implants*, 204.

Chapter 7 **Politics and Medical Progress**

1. See Nicholas B. King, "Security, Disease, Commerce: Ideologies of Postcolonial Global Health," *Social Studies of Science* 32 (2006): 763–89.
2. Data are from the ISI Web of Knowledge™. Overall, from 1984 to 2007, 3,552 papers are listed. Of these, 75 percent came from the United States (46 percent), Germany, England, and Australia.
3. M. P. Haggard, *First European Symposium on Paediatric Cochlear Implantation*, abstract 63 (Nottingham, September 1992).
4. R. Carter and D. Hailey, "Economic Evaluation of the Cochlear Implant," *International Journal of Technology Assessment in Health Care* 15 (1999): 520–30.
5. Many quality of life scales have been developed. Some of them depend on inviting a group of respondents to indicate how they view the relative burden of various limiting or discomforting conditions. In Carter and Hailey's study, "Decisions on the appropriate dimensions to include—the pre/postimplantation values for each dimension for three classes of implantees (profoundly deafened adults, partially deafened adults, and children)—reflected the judgment of a doctor experienced in clinical aspects of cochlear implantation and of a researcher in the routine use and outcomes of the technology . . . their judgments were combined into one set of results."
6. A. K. Cheng et al., "Cost-Utility Analysis of the Cochlear Implant in Children," *Journal of the American Medical Association* 284 (2000): 850–56.
7. See S. M. Garber, S. Ridgely, M. Bradley, and K. W. Chin, "Payment under Public and Private Insurance and Access to Cochlear Implants," *Archives of Otolaryngology: Head and Neck Surgery* 128 (2002): 1145–52. This study, sponsored by the Advanced Bionics Corporation and carried out by RAND Health, found that under Medicare and Medicaid, hospitals lost anywhere between $5,000 and $20,000 per implant. Even private health insurance often failed to cover total costs.
8. Of the 3,500+ articles generated by the Web of Knowledge™ with the search term "cochlear implant," 104 articles also include the term "quality of life." Only 4 of these had appeared prior to 1995, and 50 percent had been published since 2005. Searching within the 3,500+ articles with the term "school" yields 80 publications, of which 60 have appeared since 2000 (numbers are as of November 11, 2008).
9. Professionals elsewhere have not been much impressed by this work. A major paper reporting its findings (Gunilla Preisler, Anna-Lena Tvingstedt, and Margareta Ahlström, "A Psychosocial Follow-up Study of Deaf Preschool Children Using Cochlear Implants," *Childcare Health and Development* 28 (2002): 403–18) had received, as of November 2008, just ten citations, of which only three were from medical journals. Scores of other publications from the same year, reporting technical or surgical developments or using more orthodox outcome measures, had received thirty to sixty or more citations.
10. M. C. Creditor and J. B. Garrett, "The Information Base for Diffusion of Technology: Computed Tomography Scanning," *New England Journal of Medicine* 297 (1977): 49- 53.

11. Perhaps technologies of organ transplantation, offering hope of life to people suffering from renal or heart failure, provide the most dramatic example of this. As Berkeley anthropologist Nancy Scheper-Hughes has so powerfully documented, organ transplantation has come to mean very different things in the North and in the South, among the rich and the poor, the white and the black people of the world. The procedure offers hope to the affluent sick in some places; in others, it provokes fear among the poor and exploited (that their organs will be stolen from them). Though in this case inflated by violence and criminality, the global economy of hope and fear showed by organ transplantation in some respects parallels that of the cochlear implant. See Nancy Scheper-Hughes, "The Global Traffic in Human Organs," *Current Anthropology* 41 (2000): 191–224.
12. Thomas Parke Hughes, *Networks of Power: Electrification in Western Society, 1880–1930* (Baltimore: Johns Hopkins University Press, 1983); Thomas Parke Hughes, "Technological Momentum," in Merritt Roe Smith and Leo Marx, eds., *Does Technology Drive History? The Dilemma of Technological Determinism* (Cambridge: MIT Press, 1994), 101–13.
13. Crouch, "Letting the Deaf Be Deaf."
14. I am not discussing here another line of critique based on the idea of "U.S.-style bioethics as an imperialist project." On this, see Fox and Swazey, *Observing Bioethics*, especially chapter 10, "The Development of Bioethics in the Islamic Republic of Pakistan."
15. N. Daniels, "Equity and Population Health: Towards a Broader Bioethics Agenda," *Hastings Center Report* 36 (2006): 22–35. See also Fox and Swazey's interview with Norman Daniels in *Observing Bioethics*, 186–87.
16. Gregg M. Bloche, "Clinical Loyalties and the Social Purposes of Medicine," *Journal of the American Medical Association* 280 (1999): 268–74.
17. That the demands of a state can be immoral is exemplified by the many examples of medical complicity with state-instigated human rights abuses. See Gregg M. Bloche, "Caretakers and Collaborators," *Cambridge Quarterly of Healthcare Ethics* 10 (2001): 275–84.
18. Margaret Olivia Little, "Cosmetic Surgery, Suspect Norms, and the Ethics of Complicity," in Erik Parens, ed., *Enhancing Human Traits: Ethical and Social Implications* (Washington, DC: Georgetown University Press, 1998), 162–76.
19. Charles Taylor, *Multiculturalism and "The Politics of Recognition"* (Princeton, NJ: Princeton University Press, 1992).
20. For a subtle and comprehensive discussion of the notion of collective rights, see Kwame Anthony Appiah, *The Ethics of Identity* (Princeton, NJ: Princeton University Press, 2005).
21. Difficulty of access to media and to political decision makers can be seen as an indicator of injustice. See Iris Marion Young, "Activist Challenges to Deliberative Democracy," *Political Theory* 29 (2001): 670–90.
22. See, for example, Michael Pursglove and Anna Komarova, "The Changing World of the Russian Deaf Community," in Monaghan et al., *Many Ways to Be Deaf*, 249–59.
23. Pursglove and Komarova write that "according to one expert, the term *deaf culture* in its Western sense, is understood by no more than 20 people." Pursglove and Komarova, "The Changing World," 255.
24. Senghas, "New Ways to Be Deaf in Nicaragua," 261.
25. Fox and Swazey suggest that in the last few years this fate has befallen bioethics in the United States, as the field and its institutions has split between "liberals" and "conservatives." See Fox and Swazey, *Observing Bioethics*, chapter 11.

26. For a discussion of the British National Institute of Clinical Excellence (NICE) in these terms, see Celia Davies, "Grounding Governance in Dialogue? Discourse, Practice, and the Potential for a New Public Sector Organizational Form in Britain," *Public Administration* 85 (2007): 47–66.
27. The panel also included one psychologist, one educator of the deaf, one medical ethicist, and one representative of the general public.
28. National Institutes of Health, *Consensus Development Conference Statement Cochlear Implants in Adults and Children* 13, no. 2 (May 15–17, 1995), 4.
29. See http://www.ccne-ethique.fr (consulted November 5, 2008).
30. Young, "Activist Challenges." There is considerable debate among political theorists regarding the nature of deliberative democracy, including, for example, the extent to which considerations of process and substance can be separated. For an alternative approach to Young's, see Amy Gutmann and Dennis Thompson, "Deliberative Democracy beyond Process," *Journal of Political Philosophy* 10 (2002): 153–74.
31. I would like to think that my attempt at intervening had some effect. As exemplified by the stakeholder meeting that took place in 1995 and the platform that was then established, dialogue did indeed take place in the Netherlands. But looking back, it is clear that its effects on policy, practice, and the demands of parents were brief and ultimately inconsequential. My ethnographic intervention too was unable to "hold the line." Young's "activist" would again find her position vindicated.
32. Christiansen and Leigh, *Cochlear Implants*, 321.
33. I. Sinha, L. Jones, R. L. Smyth, and P. R. Williamson, "A Systematic Review of Studies That Aim to Determine which Outcomes to Measure in Clinical Trials in Children," *PLOS Medicine* 5 (2008): 0569–0578. I am grateful to Elizabeth Vroom for bringing this article to my attention.
34. Alberto Melucci and Leonardo Avritzer, "Complexity, Cultural Pluralism, and Democracy: Collective Action in the Public Space," *Social Science Information* 39 (2000): 507–27, 521.
35. Brown, Zavestoski, et al., "Embodied Health Movements," 72.
36. Rabeharisoa, "From Representation to Mediation," 504.
37. Brown, Zavestoski, et al. make a similar point when they point out that "illnesses characterized by diffuse symptoms . . . are less likely to see the emergence of strong [health movements] than medically accepted diseases . . . without some formal recognition from scientific and medical professionals the gatekeepers of government money are not likely to see a need for funding research," Brown, Zavestoski, et al., "Embodied Health Movements," 73–74.

Index

About the Author

STUART BLUME is emeritus professor of science dynamics at the University of Amsterdam (the Netherlands). Educated at Oxford University, he worked at the University of Sussex, the London School of Economics, and in various British government departments before moving to Amsterdam. He chairs the Innovia Foundation for Medicine, Technology, and Society, which he helped establish in 2000. Publications include *Toward a Political Sociology of Science* and *Insight and Industry: The Dynamics of Technological Change in Medicine*.